Natural Gas

Its Role and Potential
in Economic Development

Natural Gas

Its Role and Potential in Economic Development

EDITED BY

**Walter Vergara,
Nelson E. Hay
and Carl W. Hall**

Routledge
Taylor & Francis Group

LONDON AND NEW YORK

First published 1990 by Westview Press

Published 2018 by Routledge
52 Vanderbilt Avenue, New York, NY 10017
2 Park Square, Milton Park, Abingdon, Oxon OX14 4RN

Routledge is an imprint of the Taylor & Francis Group, an informa business

Library of Congress Cataloging-in-Publication Data
Natural gas: its role and potential in economic development/edited
by Walter Vergara, Nelson Hay, Carl Hall.
    p. cm.—(Westview special studies in natural resources and
energy management)
    Includes index.
    ISBN 0-8133-7812-5
    1. Gas industry—Developing countries. 2. Gas, Natural.
3. Economic development. I. Vergara, Walter, 1950–   . II. Hay,
Nelson E. III. Hall, Carl W. IV. Series.
HD9581.D442N38   1990
338.2′7285′091724—dc20   89-33756
                                                                    CIP

ISBN 13: 978-0-367-01331-8 (hbk)

ISBN 13: 978-0-367-16318-1 (pbk)

Contents

Preface

This book is concerned with the multifaceted character of natural gas and the potential for its utilization as a powerful tool for development during the next century. This remarkable natural resource is indeed an abundant, clean, yet so far underutilized energy and chemical feedstock. For decades, natural gas was viewed as the "unwanted child of oil exploration": a nuisance whose disposal required the diversion of resources, a material that in many fields only found use as a reinjection medium for the production of additional oil. Gas's relative position has now drastically changed in reference to competing resources and is today being viewed as an "energy feedstock and raw material of choice." Many factors have contributed to this new outlook.

The main purpose of the book is to shed light on the economic, environmental and institutional factors that are key for gas to capture its huge potential in energy and industrial applications worldwide and to review the current situation of specific gas uses in the power, fertilizer, chemical and transportation sectors.

The book shows that the natural gas base is abundant, far superior to competing hydrocarbon resources and better distributed. Proven gas reserves are now estimated at over 11,000 exajoules (10,900 Quads): in excess of 175 years of reserves at the current rates of consumption and better distributed than oil. Over 100 countries on all five continents now have substantial volumes of gas reserves. It is also shown that gas enjoys a clear environmental advantage over other hydrocarbon resources in most applications and has the potential to contribute to the alleviation of some of the most pervasive global environmental problems. Gas could be instrumental in the reduction of the rate of growth of carbon dioxide emissions, dent the rate of formation of sulphur dioxides and improve the thermal efficiency of power applications.

This book also establishes that the key obstacles to increased gas utilization include economic factors, with ignorance of gas valuation and pricing often being central to the lack of appreciation of the economic advantages of gas utilization. The book addresses the issues of estimation of gas costs and pricing of natural gas to provide for sufficient incentives for producers and users. Likewise, institutional and policy factors are shown to have played an important role in the delays experienced in gas development. With natural gas growth likely to take place mostly in developing countries, the policy issues and the institutional basis for gas development acquire an additional urgency; these include a consensus on the desirable role of gas on the overall energy economy, the choice of markets, the financing of gas development and the pricing of gas.

As a complement to the analysis of economic, environmental and institutional and policy issues in gas development, the book also deals with the specifics of gas uses in a number of sectors chosen for their relevance in the development of industrial economies. The book examines the use of gas in the power sector. It analyzes the potential role of natural gas-fired gas turbines in the current revolution being experienced in electricity-generating technology. Through the use of gas turbines, much needed relief may be at hand for central utilities in developing and industrialized countries facing the massive capital requirements for the installation of central hydro or thermal units. The text examines the technological innovations that have enabled gas turbines to become competitive in cogeneration and asserts that these innovations, the bullish outlook for competitively-priced natural gas reserves and the environmental advantages of gas contribute to an optimistic outlook for widespread use of gas turbines in power applications.

The book also establishes that availability of gas and natural gas liquids have had a lasting influence in the current potential and outlook for synthetic materials produced by the chemical industry. The widespread availablility of gas and gas liquids has been instrumental in the emergence of new and established producers in the global market of chemicals and contributed to the increase in competitiveness of synthetic materials in the world economy. The chemical sector is at the beginning of an evolutionary track that has gradually enabled petrochemicals to compete and outperform traditional materials. Gas is expected to continue to play a substantive role in the realization of this potential.

The traditional use of gas in the fertilizer industry is also examined. The industry is found currently to favor the use of gas as feedstock over competing resources because of energy and economic considerations. The widespread availability of gas is shown to be a key factor behind the improved production of fertilizers by major food producers and likely to continue to support the future requirements for nitrogen-based fertilizers.

Nowhere has gas potential been more misunderstood than in the transportation sector. The chemical characteristics of natural gas and the nature of the internal-combustion engine combine to make properly conditioned gas an ideal fuel for transportation. Yet, with the exception of a few notable cases, gas has played only a marginal role in the transportation market. The dominant role of gasoline and diesel and the inertia of powerful economic trends in the sector have also delayed the use of gas in transportation. This market is possibly the largest single application for natural gas in the next century. The book describes the painful historical development of gas use in transportation and outlines the potential and obstacles for its widespread application. The book also reviews the existing synergy between gas development and the development of alternative gaseous fuels.

Walter Vergara
Nelson E. Hay
Carl W. Hall
Washington, D.C.

1

The Emergence of
Natural Gas

Nelson E. Hay

Natural gas is an abundant resource with broad worldwide distribution. Yet, given its inherent economic and environmental advantages, it may be considered underutilized. This situation is now changing dramatically. Natural gas is emerging as a "fuel and raw material of choice" for coming decades, with important implications from a development perspective.

Certainly the natural gas resource base is now recognized to be large. The economically recoverable natural gas resource of 10,869 quadrillion Btu (quads) (11,471 exajoules-EJ) represents 175 years of supply at today's consumption rate.[1] The International Gas Union (IGU) projects that by 2020, 74-90 years of economically recoverable gas resources at 2020 consumption rates will remain, even if maximum potential gas demand rates are attained.[2] North America and Western Europe would be the only areas where economically recoverable natural gas resources of less than 50 years would remain in 2020 at maximum demand rates.[3] The total in-place worldwide natural gas resource base, including sources not yet economically producible with today's technology, is several orders of magnitude greater.[4] Clearly, there are sufficient gas resources in place worldwide to satisfy any plausible requirements through the twenty-first century.

These resources have not gone totally unnoticed. World gas consumption increased by 73.8 percent between 1970 and 1987, an average increase of 3.3 percent annually. Excluding the United States, where natural gas use decreased by 20.6 percent over the period, natural gas consumption in the rest of the world rose by 220 percent.[5] This broad growth in natural gas use was based on plentiful supply, moderate costs, environmental benefits and national programs to reduce dependence on oil.

In spite of this increased natural gas consumption, the natural gas resource base remains far less utilized than the oil resource base. In every major oil and gas producing country the reserves to production ratio (R/P) is higher for gas than for oil. Worldwide the R/P ratio was 58.3/1 for gas and 41.5/1 for oil in 1987.[5] Why is this so?

There are several answers. Initially, natural gas production was an unwanted byproduct of oil production. We really didn't realize how much of it there was (and is) from a resource perspective. With time and technological development (e.g. deep drilling, offshore drilling) we are coming to recognize that the natural gas resource is large, as we continue to find surprising amounts of it. Major oil and gas producing companies of the past are now saying that they will be gas and oil producers in the future -- gas will come to have the greater importance.

Even in those instances where we knew we had a lot of gas, development of a transportation infrastructure faced significant hurdles. From an international trade perspective, an economically viable seaborne means of moving gas did not exist until development of modern liquefied natural gas (LNG) technology. Significant pipeline and local distribution infrastructures developed in the major industrialized nations, but in much of the rest of the world, particularly the third world, the conventional wisdom was (and has continued to be) that economics favored the development of an electricity infrastructure *in lieu* of a natural gas infrastructure. In other words, it has been believed that many nations cannot afford to develop both natural gas and electricity infrastructures, but must choose one. Because everyone needs electricity for lighting, communications, etc., the obvious choice has been electricity, although electricity is clearly a less efficient and more polluting use of resources in many applications.

Today, the basic underpinning of this "electricity only" theory has collapsed as a result of developments in electricity generation technology which eliminate the need for large, central station electricity generating plants. The "electricity only" theory depended upon the fact that electricity generation was highly capital intensive and buildable only in large increments. Available technologies required large, central station generating plants in order to optimize plant economics. The necessity to optimize plant economics, and the necessity of building these very large, very expensive plants in turn necessitated maximizing electricity demand to pay for the plant.

As discussed in Chapters 3 and 5, the successful commercialization of natural gas combined cycle and cogeneration turbine technologies have revolutionized electricity economics. With these technologies, and others which are emerging, dispersed electricity generating capacity can be added in smaller, more flexible increments (e.g., under 100 MW), in a fraction of the time, and at one third to one fifth the capital cost -- resulting in lower cost electricity and dramatically reduced environmental impact (see

2

Chapter 5). Thus, the need to support a one-time, massive investment in central station electricity generating capacity no longer exists where natural gas is available.

Once natural gas production and pipeline capacity is being installed to service electricity generation (or other large-scale uses such as ammonia or methanol production), the next logical step is to hook up other users along the pipeline to natural gas as well. Initially these might be large industrial and commercial users and fleets of vehicles (as discussed in Chapter 5), but the establishment of residential and commercial grids would logically follow.

This brings us to another key element of the "electricity only" development theory -- namely, that it is not economically, technically or even politically practical to install natural gas main and to retrofit homes and other buildings to use gas in existing urban areas and communities. With the advent of plastic and flexible piping, this notion is now being disproven in countries as diverse as Egypt, Barbados and Brazil.

The past decade has seen an unprecedented outpouring of new natural gas technology. With today's natural gas end-use technology (and more so with the coming generation, e.g. fuel cells; self-fueling heating and cooling; advanced, factory-built vehicles) direct use of natural gas or cogeneration with natural gas is the most efficient, least polluting means of providing energy. Where the high cost of electricity is a serious problem, use of gas to generate the electricity can often cut the cost by as much as half; and direct use of gas can often cut energy costs by two-thirds. Where the high cost of emissions controls are an obstacle to their use, few controls are required with natural gas.

Changing perceptions with regard to environmental protection, as discussed in Chapter 3, are the final blow to the "electricity only" theory. Until recently it was widely believed that developing nations couldn't afford environmental protection. This argument weakened at the international institution level with increasing recognition of the trans-national and even worldwide nature of problems such as acid rain and climate change. Now, increasingly, protecting natural resources is seen at the developing country level as a necessary survival strategy rather than a luxury, a means of sustaining progress rather than constraining it. Pollution control can be very expensive. That natural gas offers a low cost means of minimizing pollution while also choosing a least-cost energy strategy is a benefit whose time has come.

Overview of
Natural Gas Resources

The terminology used in reporting gas resources can be quite confusing. Generally, there are three basic levels: *proved reserves*; economically *recoverable resources*; and *in-place* or *total resources*. Gas in the first category already has been discovered, and is considered producible under current economic and operating conditions. Gas in the second category is believed to exist and is estimated to be producible under economic and operating conditions which are presently considered likely to exist in the foreseeable future. Gas in the third category is the total resource believed to exist in all deposits, of which some portion (depending on technology and economics) could be produced. With time and technological development more and more of this gas resource will become economically recoverable.[4]

Cutting across the categories just defined above, gas resources are often also broken out into conventional and unconventional or non-conventional. The term "non-conventional," for the purposes of this chapter, refers to resources such as tight sands, Devonian shale, coal seam methane and other gas resources requiring enhanced gas recovery techniques. The existence of these sources is well documented, and estimates are based on numerous known occurrences worldwide.

There are, however, two potential sources of gas which differ from the other non-conventional sources because of their intangible nature and the vast quantities of energy which could result if theories supporting their existence are proven to be correct. These are methane from gas-hydrates and abiogenetic methane or deep-earth gas. While the existence of gas-hydrates is acknowledged, very little is known about this potential gas source. The theories surrounding deep-earth gas are at present unproven. As more knowledge of these sources is gained, there is the potential for extremely large quantities of gas being made available worldwide.

A more detailed discussion of natural gas resources and production technology appears in Appendixes A and B. The key point to be made here is that when all gas categories are considered, and compared to current consumption, there are clearly sufficient gas resources in place worldwide to satisfy any reasonable demand requirements through the twenty-first century and beyond (see Table 1.1). Further, it is widely recognized that many known gas reservoirs are currently untapped, often in nations which rely heavily upon imported petroleum or refined petroleum products.

Table 1.1.   World Gas Consumption vs. Resources

	1986[1] Consumption	Proved[1] Reserves	Remaining[1] Recoverable	In-Place Resource[4] (Incl. Non-Conven).	In-Place Resource[4] (Incl. Speculative Non-Conven.)
Quads	62.1	3,623	7,246	143,000	271,000,000
EJ	65.5	3,826	7,645	151,000	286,000,000
BCM	1,790	104,440	208,710	4,110,000	7,810,000,000

Notes: 1 EJ = 27.3 BCM = .9478 Quads
(Quadrillion British Thermal Units = 10^{15} Btu = "Quads")
(Exajoules = 10^{18} Joules = "EJ")
(Billion Cubic Meters = 10^{9} CBM = "BCM")

Comparison of Resource Definitions
for Various Fuels

The confusion inherent in resource terminology is not limited to natural gas. It is worth noting that in 1987 the Institute of Energy Resource Studies at the Colorado School of Mines issued an important study concerning resource definitions of selected fuels.[6] The Colorado School of Mines study points out the major differences in the way coal and natural gas reserves and resources are calculated and defined. It is apparent that the existing systems are inconsistent.

In particular, the phrase "reserve base" includes subeconomic resources when used with coal. As discussed above, the term *proved reserves* is most commonly applied for those natural gas resources which have been confirmed to exist (as a result of drilling) and are currently producible under contemporary technical and economic conditions.

In the jargon of coal only the term "identified, economic coal reserves," would be comparable. The usual references to the "coal reserve base" describe a quantity of coal that is thought to be in place, but it does not address the amount of that coal that can be economically recovered. Thus, estimates of "the coal reserve base"   are more like the gas "in-place" than it is like gas "proved reserves" (which is limited to economically recoverable reserves). As a result, for long range energy planning issues, there is *no certainty* regarding the true relative sizes of the resource of natural gas as compared with coal.

Table 1.2. International Gas Union (IGU) World Natural Gas Reserves as of January 1, 1986[1]

	Proved Recoverable Reserves		Additional Recoverable Reserves		Total Reserves	
	$10^9 \, m^3$	%	$10^9 \, m^3$	%	$10^9 \, m^3$	%
Africa	5,860	6	10,200	5	16,060	5
Asia	6,400	6	12,600	6	19,000	6
East Europe	42,250	40	120,930	58	163,180	52
JANZ	1,680	2	1,280	1	2,960	1
Latin America	7,120	7	7,480	4	14,600	5
Middle East	26,150	25	21,850	10	48,000	15
North America	8,250	8	30,170	14	38,420	12
West Europe	6,730	6	4,200	2	10,930	4
World Total	**104,440**	**100**	**208,710**	**100**	**313,150**	**100**

Note: JANZ: Japan, Australia, New Zealand.

Regional Distribution of Recoverable Gas Resources

As shown in Table 1.2, significant, recoverable gas resources are found in all regions of the world. While the United States, Eastern Europe (including the USSR) and the Middle East account for 79 percent of reported reserves, these figures tend to be a function of the intensity with which a given region has been explored for gas. In the view of the author, natural gas reserves are probably understated on a worldwide basis. Certainly past experience would indicate this to be the case in the less explored areas. Thus, the apparent concentration of resources in certain regions may ultimately prove misleading.

World Natural Gas Consumption and Production Trends

Growth trends in natural gas consumption that typified the 1970s were slowed by the worldwide recession of the early 1980s, but have since regained strength. Earlier studies showed a worldwide movement toward increased consumption of natural gas -- both steadily increasing volumes and increasing share of total energy consumed. More recent statistics indicate that this trend continues, and the statistics reflect increased natural gas use and record natural gas consumption levels in many parts of the world.

World gas usage patterns between 1970 and 1987 show a decided movement toward increased use of natural gas. Some key features of this trend are:

- Between 1970 and 1987, gas consumption rose in all of the major industrial nations of the world except for the United States. World gas consumption rose at an average annual rate of 3.3 percent from 1970 through 1987. Between 1970-87 gas consumption by major industrial countries (excluding the US) tripled -- while the US consumption of gas declined 20.6 percent.

- Although the recession of the early 1980's temporarily slowed growth in gas consumption, especially in the western industrialized nations, statistics for 1987 suggest that the earlier overall growth trend has been re-established. This was particularly true for the nations of Western Europe which established a 6.7 percent annual growth rate for gas consumption from 1970 through 1987. More recently, from 1980 through 1987 gas consumption levels in Western Europe grew at an average rate of 1.9 percent per year.

- Of the world's industrial nations, the greatest absolute growth in gas consumption from 1970 to 1987 was in the USSR which increased from 6.2 quads in 1970 to 21.0 quads in 1987 with an average growth rate of 7.5 percent per year. In addition, the USSR has increased its exports to Europe by 46 percent from 1980 to 1987.

- Dominance of world gas statistics by the United States ended, at least temporarily, in the early 1980's. In the important categories of proved reserves, annual production and exports, the USSR now ranks first.

- Market share analysis from 1970 to 1987 confirms the growing importance of natural gas in the USSR; gas increased from 21 percent to 36 percent of that country's total energy supply. A similar analysis showed reduced gas market share from 34 percent to 24 percent in the US over the same period, although many analysts believe that this trend is now reversing thanks to an improved regulatory environment and new gas technologies.

Table 1.3. Natural Gas Consumption (Trillion Btu)

Country	1970 Consump-tion	Share	1980 Consump-tion	Share	1985 Consump-tion	Share	1987 Consump-tion	Share
Australia/ New Zealand	59	.03	396	.12	634	.17	731	.18
Austria	98	.12	169	.16	184	.18	184	.17
Belgium/ Luxembourg	164	.09	418	.21	338	.18	346	.17
Canada	1,275	.20	1,994	.22	1,811	.20	1,667	.18
China	129	.01	472	.02	464	.02	518	.02
France	363	.06	954	.13	943	.13	1,008	.13
Germany(FRG)	507	.05	1,796	.17	1,663	.16	1,796	.17
Italy	480	.10	925	.16	1,102	.20	1,303	.22
Japan	140	.01	947	.07	1,451	.10	1,472	.10
Netherlands	745	.32	1,224	.41	1,314	.47	1,372	.46
Spain	4	-	72	.02	86	.03	112	.04
UK	437	.05	1,674	.21	1,958	.24	2,023	.25
US	22,000	.34	20,513	.28	18,194	.25	17,471	.24
USSR	6,201	.21	13,266	.28	19,199	.35	21,042	.36
Western Europe	2,798	.07	7,319	.15	7,765	.15	8,366	.16
World	36,219	.18	52,906	.19	60,433	.20	62,939	.20

Notes: Share is the percentage of gas consumption to total energy consumption in the particular country or region. 1 trillion Btu = .00106 exajoules.

Source: *BP Review of World Gas*, July 1988.

Consumption Trends

In the seventeen-year period from 1970 to 1987, natural gas consumption rose in all the countries shown in Table 1.3, with the sole exception of the US. In the U.S. gas use decreased 20.6 percent. In the rest of the world gas consumption rose by 220 percent. Western Europe increased gas consumption at an average annual rate of 6.7 percent. Most of this growth occurred prior to 1980; in the 1980 to 1987 time frame this growth rate for Western Europe was only 1.9 percent per annum. The USSR, now the largest national consumer of natural gas, increased consumption by 239 percent (7.5 percent per annum) from 1970 to 1987. Gas consumption increased over this time period in the USSR from 6.2 quads to 21.0 quads.

Overall, world gas consumption increased 73.8 percent (3.3 percent per annum). These substantial increases in natural gas use indicate that both supplies of natural gas and the infrastructure necessary to transmit and consume gas were expanding rapidly.

Most nations have developed their gas resources intensively over the past fifteen years. The quantities of gas and oil proved reserves at year-end 1987 are shown in Table 1.4. In total, world gas reserves, as reported by British Petroleum's *BP Review of World Gas*,[5] were 80 percent as large as world oil reserves. Five countries held 65 percent of the world's natural gas proved reserves - USSR (38 percent), Iran (13 percent), US (5 percent), Abu Dhabi (5 percent) and Qatar (4 percent). Saudi Arabia, Norway and Canada accounted for an additional nine percent of the world reserves. These eight countries hold 74 percent of the proved reserves of natural gas.

The reserves to production (R/P) ratios shown in Table 1.4 are a measure of how rapidly these reserves are being produced. For gas, the US and UK have the lowest R/P ratios indicating that the gas reserves are being produced more rapidly. Similarly, the R/P ratios for oil show the US and UK as having ratios less than ten. In contrast, Saudi Arabia, Kuwait, Iraq, Iran and Abu Dhabi report R/P ratios for oil of over 100. In every major oil and natural gas producing country, the R/P ratio was higher for gas than oil. Worldwide the R/P ratio was 58.3/1 for gas and 41.5/1 for oil in 1987.

The reserves ratios of gas to oil, as listed in Table 1.4, are also indicative of the proved reserves available for use in the nations listed. Six of the nations listed (Netherlands, Qatar, USSR, Canada, Norway and the US) have more energy in proved gas reserves than in oil.

The statistics of Tables 1.3 and 1.5 show that worldwide gas consumption and production have increased dramatically since 1975. The average growth rates during this period moderated from 1980 to 1986 as compared to the early years of this period. Nevertheless, it would appear that 1987 data generally indicate a resurgence in gas use and a return to the longer term growth trend.

On a regional basis North America showed a reduction in gas consumption since 1970. The major factor was that US consumption has been declining since the early 1970s; but also Canadian gas consumption declined in 1985-86-87 from the levels achieved in 1980.

9

Table 1.4. Proved Reserves of Natural Gas and Oil (Quadrillion Btu)

Country	Natural Gas Gas Reserves	R/P*	Oil Oil Reserves	R/P*	Reserves Ratio Natural Gas/Oil
Abu Dhabi	187	-**	484	-**	0.39
Canada	83	38.8	40	11.3	2.08
China	32	63.1**	96	18.2**	0.33
Iran	500	-**	508	-**	0.98
Iraq	25	-**	536	-**	0.05
Kuwait	36	-**	508***	-	0.07
Netherlands	65	29.0	-	-	(very large)
Norway	108	-**	76	-**	1.42
Qatar	158	-**	16	-**	9.88
Saudi Arabia	144	-**	908	-**	0.16
UK	22	14.9	28	5.5	0.79
US⁴	191	11.5	168	9.0	1.14
USSR	1480	56.6	320	12.9	4.63
Regional					
Africa	252	-**	296	29.4	0.85
Asia	205	60.5	92	12.1	2.23
Australasia	22	34.7	8	7.7	2.75
Centrally Planned Economies	1541	52.9	424	13.7	3.63
Latin America	234	75.0	644	49.3**	0.36
Middle East	1105	-**	3060	-**	0.36
North America	292	15.0	208	9.4	1.40
Western Europe	223	34.7	116	14.6	1.92
Total World	**3874**	**58.3**	**4848**	**41.5**	**0.80**

* Reserves to production ratio for 1987.

** Over 100 Years.

*** Less than 2 Quads.

Note: 1 quadrillion Btu = 1.06 exajoules.

Source: *BP Review of World Gas*, July 1988.

In Western Europe the trend of rising gas consumption (1970-1980) also changed in 1981-1983 when a temporary decline occurred (not shown in Tables 1.3 and 1.5). By 1985, however the strong growth pattern in gas use was again evident. Only Belgium and Luxembourg in this region

Table 1.5. Natural Gas Production (Trillion Btu)

Country	1975	1980	1985	1986	1987	Percent of World Production 1987
Algeria	344	781	1,361	1,278	1,516	2
Australia/ New Zealand	220	439	619	706	709	1
Canada	2,756	2,794	2,786	2,599	2,747	4
China	356	493	464	475	500	1
France	276	284	184	140	130	-
Germany (FRG)	548	569	493	554	6341	
Indonesia	84	676	1,105	1,148	1,184	2
Italy	524	418	500	583	594	1
Japan	88	79	79	76	79	-
Mexico	596	1,102	983	983	1,004	1
Netherlands	2,748	2,786	2,606	2,272	2,282	3
Norway	8	954	922	983	1,069	2
UK	1,276	1,296	1,501	1,577	1,573	2
US	19,626	20,023	16,888	16,488	16,798	25
USSR[1]	10,416	15,379	23,407	24,973	26,417	39
Venezuela	460	608	626	634	698	1
World	45,276	54,407	62,381	63,860	67,194	100

Note: 1 trillion Btu = .00106 exajoules.
Source: *BP Review of World Gas*, July 1988.

reported a 1987 level of gas consumption below the 1980 levels. The other Western European nations generally increased gas consumption in 1987 well above previous levels. In fact, Western European nations as a group increased gas consumption 14 percent in the 1980-1987 period due in large measure to increased imports from the USSR.

For that reason, it is interesting to contrast the Western European gas statistics with those of the USSR.[7] The Soviet policy of increased gas production and sales clearly offset any influence of slower international economic growth in the early 1980's. In fact, Soviet gas consumption continued to increase decisively, and it is clear that the other Eastern European countries (listed in Table 1.6) continued to expand their use of gas rapidly.

Table 1.6. Gas Imports and Exports (Trillion (10^{12}) Btu)

Country	1970	1980	1985	1986	1987	Imports as % of Total Gas Consumption 1986
Austria	38	108	137	139	133	74
Belgium & Luxembourg	150	394	335	295	338	99
Bulgaria	-	-	184	189	223	92[2]
Czechoslovakia	-	270	378	414	428	95
France	125	718	893	626	968	97
Germany (GDR)	-	199	252	256	263	65[2]
Germany (FDR)	125	1,436	1,274	1,318	1,609	91
Hungary	-	133	155	166	176	45[2]
Italy	-	519	691	706	839	65
Poland	-	161	194	230	266	64[2]
Netherlands	-	126	68	61	68	5
Japan	-	-	1,357	1,390	1,444	99
Spain	3	75	86	90	90	80
UK	33	397	461	454	443	22
US	846	1,010	828	770	904	5
USSR	-	143	86	79	79	-
Yugoslavia	-	49	137	137	187	69[2]
Pipeline Exporters (in 10^{12} Btu)						
Afghanistan	N/A	N/A	N/A	N/A	79	N/A
Algeria	N/A	N/A	N/A	N/A	403	N/A
Bolivia	N/A	N/A	N/A	N/A	79	N/A
Canada	779	796	803	770	904	55[1]
Denmark	N/A	N/A	N/A	N/A	29	56[1]
Netherlands	378	1,823	1,267	1,084	1,256	93[1]
Norway	N/A	890	950	961	1,055	1,864[1,2]
USSR	N/A	2,0693	2,437	2,768	3,031	15
Total	N/A	N/A	N/A	N/A	6,836	111

(Table 1.6 cont'd. next page.)

Table 1.6. Gas Imports and Exports (Trillion (10^{12}) Btu) (Cont'd.)

Country	1970	1980	1985	1986	1987
LNG Exporters (in 10^{12} Btu)					
Abu Dhabi	N/A	-	108	108	104
Algeria	N/A	216	443	432	482
Brunei	N/A	N/A	252	252	256
Indonesia	N/A	438	738	738	763
Libya	N/A	N/A	43	32	29
Malaysia	N/A	N/A	209	245	292
US	44	45	50	50	47
Total	108	1,116	1,843	1,858	1,973

[1] Ratio shown is exports to consumption expressed as a percentage.

[2] Data for the latest year available -- usually 1987 taken from *Annual Bulletin of General Energy Statistics for Europe*, Economic Commission for Europe, United Nations, 1988.

[3] 1981 data from Economic Commission for Europe, Committee on Gas, "The Gas Situation in the ECE Region in 1983 and Prospects."

Sources: *Annual Bulletin of Gas Statistics*, Economic Commission for Europe, United Nations, 1988; *BP Review of World Gas*, July 1988; and *BP Review of World Gas*, September 1987.

Production Trends

World gas production has increased steadily since 1975. The average rate of increase was 3.3 percent per year from 1975 to 1987. Gas production increased most dramatically in the USSR. In 1975 the USSR production was only 10.4 quads per year. By 1987 the USSR produced 39 percent (26.4 quads) of the world's gas at a level two and one-half times the 1975 level - an annual growth rate of 8.1 percent. Available evidence of Soviet pipeline construction, field development and gas reserves indicate that this rate of increase may moderate but even now, steady increases in production capability are likely.

In the US, the national production capability has declined to a level of about 18 Tcf per year (actual production in 1987 was 16.8 quads).[8] US production capability could remain at this level indefinitely if drilling activity is strong. On the other hand, with low levels of gas well drilling such as those reported in 1986-87, some loss of production capability will continue as the producing wells mature and decrease in production.

From a world perspective in the long run, the 1987 US share of 25 percent of world production can be expected to decrease or remain constant. There is only a small chance that the future rate of the world increase in gas production will be exceeded by conventional production in the US.

Production in the western industrialized nations other than the US could be generally characterized as maximization of domestic production. Rises and falls of national gas demand are being offset by varying imported gas volumes.[9] Canada, UK, Norway and The Netherlands are the four most important western producing nations other than the US. As can be noted in Tables 1.3 and 1.4, Canadian production and consumption levels dropped slightly from 1975 to 1987 despite the fact that world consumption levels increased. In both the UK and Norway the development of North Sea resources made possible production (and for the UK consumption) level increases. In Australia/New Zealand production and consumption grew rapidly as the gas resources of this area were developed and the gas infrastructure came into being.

In France production levels declined from 1975 to 1987. This was in part a result of aging of the production wells, and so reflects a loss in national production capability which has not been offset by the drilling of new gas wells. In Italy and West Germany gas production increased by 13 percent and 16 percent respectively from 1975 to 1987.

Table 1.6 presents data on national imports and exports. Of the European nations, West Germany is the largest importer of gas with the 1,609 Btu in 1987 coming primarily from The Netherlands and USSR. The importance of gas imports to most of the countries listed in Table 1.6 is the fact that these imports represent more than two-thirds of the total importing nation's gas consumption. The dominance of the USSR as a world exporter of gas is clear. The USSR export level increased 46 percent to 3.0 quads from 1980 to 1987. Based on the infrastructure in place in the USSR, this level will probably increase in future years.

14

Table 1.7. World Natural Gas Trade (1979 - 1985)

Year	Natural Gas Trade (Bcf) Total	Pipeline	LNG	Global Natural Gas Production (Bcf)	Total Gas Trade As % of Production	Pipeline Gas Trade As % of Production	LNG Trade As % of Production
1979	6,763	5,534	1,229	51,570	13.1%	10.7%	2.4
1980	7,800	6,584	1,216	52,750	14.8%	12.5%	2.3
1981	7,144	6,034	1,110	54,160	13.2%	11.1%	2.0
1982	6,745	5,508	1,237	53,970	12.5%	10.2%	2.3
1983	6,965	5,421	1,544	53,970	12.9%	10.0%	2.9
1984	7,630	5,875	1,755	59,030	12.9%	10.0%	3.0
1985	8,260	6,443	1,817	61,470	13.4%	10.5%	3.0
Avg. Annual Growth Rates	3.4%	2.6%	6.7%		3.0%		

Note: 1 Bcf = 28.3 million cubic meters.

International Natural Gas Trade

Total worldwide international natural gas trade, including CNG and liquefied natural gas (LNG), was 8,260 billion cubic feet (Bcf) in 1985. (See Table 1.7.) (1 billion cubic feet = 28.3 million cubic meters) This figure is 22.1 percent greater than the 6,763 Bcf of natural gas traded in 1979. For illustrative purposes, this would be enough gas to fuel 50 to 100 million vehicles if the entire amount were used for vehicular purposes.

Since 1979, total natural gas trade has experienced an average annual growth rate of 3.4 percent, slightly greater than the 3.0 percent growth rate in global natural gas production during the same period. Thus, traded natural gas expressed as a percentage of total worldwide production has remained relatively stable over the 1979-1985 period.

Pipeline Natural Gas Trade

International trade in pipeline gas stood at 6,443 Bcf in 1985, 16.4 percent greater than the 1979 level of 5,534 Bcf. During the intervening years, global pipeline natural gas trade peaked at a record high of 6,584 Bcf in 1980 and bottomed out at 5,421 Bcf in 1983. Traded pipeline gas expressed as a percentage of total worldwide production was 10.5 percent in 1985, slightly less than the 10.7 percent figure achieved in 1979. This percentage fell steadily during the 1980 (12.5 percent) to 1984 (10.0 percent) period and is illustrative of the slow growth in pipeline traded gas (2.6 percent growth per annum) relative to worldwide production (3.0 percent growth per annum).

Leading exporters of pipeline natural gas include (1985 data) the USSR (2,510 Bcf to Europe), the Netherlands (1,524 Bcf shipped within Western Europe), Canada (926 Bcf to the US), and Norway (907 Bcf to Western Europe). Collectively these nations ship 91.1 percent of all natural gas traded via pipeline. Major importers of natural gas via pipeline are West Germany (1,544 Bcf, of which more than half is from Norway), the United States (926 Bcf from Canada), Italy (710 Bcf from Algeria, USSR and the Netherlands), and France (626 Bcf from the Netherlands, USSR and Norway).

The outlook for global natural gas traded via pipeline is expected to remain solid as economic and environmental incentives drive gas demand higher, particularly given new developments such as the U.S.-Canada Free Trade Agreement and the proposed Qatar to Turkey to Europe pipeline.

LNG Trade

Global trade in LNG was equal to 1,817 Bcf in 1985 (by comparison, the world trade in LNG is 13 times greater than the world trade in methanol, in terms of energy content of the fuel moved). This represents a 47.8 percent gain over the 1979 level of 1,229 Bcf. The intervening years saw LNG trade bottom out at 1,110 Bcf in 1981 and then steadily grow to its historical high of 1,817 Bcf in 1986.

LNG's share of total global natural gas production has jumped from 2.0 percent in 1981 to 3.0 percent in 1985. Meanwhile, between 1981 and 1985 LNG's share of the total worldwide natural gas trade has expanded from 15.5 percent to 22.0 percent. LNG trade is growing faster (6.7 percent per annum) than both global natural gas production and trade via pipeline.

Among the major exporters of LNG are Indonesia (715 Bcf to Japan), Algeria (451 Bcf to France, Luxemburg, Spain, and the United States), Brunei and Malaysia (combined 454 Bcf to Japan), United Arab Emirates (107 Bcf to Japan) the United States (53 Bcf to Japan) and Libya (37 Bcf to Spain and Italy). The major importer of LNG is Japan with 1,329 Bcf (73.1 percent of all LNG moved).

Rationale for LNG Trade. LNG offers a means of utilizing the world's large resources of safe, environmentally desirable and efficient gas energy, with potential benefits for both consumer and producer nations.

16

First, as discussed in Chapter 3 and elsewhere, gas energy is both environmentally acceptable and resource and capital efficient. LNG captures these attributes. For example, water pollution, waste consumption, and solid wastes resulting from the LNG cycle are negligible; and nitrogen oxides emissions -- the major type of air emissions from the LNG production, transportation and use cycle -- total less than 6 percent of the nitrogen oxides emissions from the coal electric cycle and 50 percent of those from the oil cycle including end-use burning.[11]

Second, LNG is safe. In a quarter century of extensive worldwide experience with the commercial shipping of LNG, the industry has compiled an outstanding safety record. Tanker incidents have been rare -- the system has proven to be sound.

From the consumer nation's perspective, LNG has also offered the opportunity to diversify fuel use as well as sources of energy. LNG gas also been attractive where consumer nations have lacked domestic natural gas resources or have been brought to face inadequate domestic gas production capability to utilize existing infrastructures.

From the producer's perspective, exports of LNG, methanol and ammonia (and their derivatives, see Chapter 7) have offered foreign exchange. They have also appeared to be options which could be brought on line more quickly and at lower capital cost than the alternative of developing producer nation domestic natural gas use infrastructures (e.g., pipelines and mains, end-use equipment, industries and powerplants).

Impediments to LNG Trade. One of the principal anticipated advantages for LNG -- that projects can be brought on line quickly -- has failed to materialize, particularly with regard to projects intended to serve the US. LNG, as a new product and technology trying to enter the U.S. market in the 1970s, faced significant regulatory obstacles. Siting concerns, which did not exist only a decade or two earlier, have become important. The frequently burdensome and conflicting regulation of the US gas industry at the Federal and state levels has not been flexible enough to adapt quickly to rapidly changing market conditions.

A second, and ultimately more fundamental impediment to LNG trade, is the capital costs of liquefaction, shipping and terminalling LNG. These fixed costs (while lower than those for methanol) are several-fold greater than the comparable costs for competitive petroleum products. Consequently, where LNG has been priced to compete in end-use markets, the netbacks to producer countries have been significantly lower than for petroleum. While this has been a long-standing source of dissatisfaction on the part of some LNG exporters, the problem was exacerbated in the mid-1980s by falling oil prices. The 30 percent decline which occurred in the wellhead price of oil meant as much as a 70 percent decline in the potential producer netback from LNG projects.

New LNG Trade Developments. With the firming of oil prices, continued new gas discoveries and increasing desire of producers to sell long shut-in gas, several new LNG long term contracts have recently been signed, and a good number proposed. LNG seems headed into renewed growth in the 1990s, although netbacks to producer countries will be low.

Future Natural Gas Supply and Demand

At the 17th World Gas conference, held in Washington, D.C. in June 1988, the International Gas Union (IGU) released its *Report of Committee J: World Gas Supply and Demand, 1986-2020*. While the author does not agree with the projections for the U.S.A. (supply and demand too low), and considers the study quite conservative, the IGU report is overall one of the best and most comprehensive publicly available studies of potential world natural gas supply and demand. This section of the chapter is an abridged quotation of the Executive Summary of the IGU report.

Methodology of the IGU Study

The IGU study was based on a questionnaire survey within IGU member countries. In addition to the questionnaires, information from other public and private sources has been analyzed and aggregated to produce eight regional estimates of potential gas supply and potential gas demand.

Supply and demand have been considered independently of each other. This approach explains why the terms potential supply and potential demand are used in this study. In other words, no effort has been made to predict the actual level at which the gas supply and demand will be balance in each area or worldwide at any point in time. Although supply and demand have been considered separately, the methodology included sufficient guarantees to arrive at realistic estimates supported by economic data. Thus, this study analyzes the long term part gas can potentially play in meeting the world's energy needs.

One of the main effects of this methodology of separate analyses is that potential supply need not necessarily be equal to potential demand. The same does not, of course, apply to actual production and actual consumption, which on a global scale should in fact be equal.

It is therefore surprising to find the actual production and consumption figures are not in balance. Although in some regions gas trade may explain part of the difference, production figures are in general found to be higher than consumption figures. This is caused by consumption figures being built up from published data on gas consumption. The published data appears to be incomplete. The actual imbalance between production and consumption figures, therefore, is no reality. Rather it is illustrative of the incompleteness of present consumption statistics.

Gas supply and demand are influenced by many factors, among which oil prices are generally considered to be predominant. Two oil price scenarios have been postulated to provide a defined background for compatible estimates of supply and demand. In the moderate scenario a crude oil price (in 1986 US$/barrel) of $20 in 2000 and of $30 in 2020 is used. In the rapid scenario a crude oil price of $40 in 2000 and of $55 in 2020 is used. In general the moderate scenario is used as the base case, whereas the rapid scenario is used as a kind of sensitivity analysis.

General Conclusions of the IGU Study

Gas Supply. The potential supply of gas in the world is estimated to increase from the 1986 level of 65.5 EJ to a 2000 level of 92-101 EJ.

During the period up to 2020 the supply potential will further increase to some 101-121 EJ. The impact of higher oil prices on gas supply is larger than the impact on gas demand.

The present proved recoverable reserves -- on a global scale -- are sufficient to cover the total potential gas production well beyond the study horizon. The additional recoverable reserves will even strengthen this position.

Gas Demand. The study shows that the gas demand outlook is different for the various regions in the world. On a world scale the potential gas demand is estimated to increase from a 1986 level of 62.7 EJ to almost 90 EJ in 2000. The high growth figures for gas demand achieved in the past are expected to be realized up to 2000, but to level off after 2000, resulting in a potential gas demand of just over 100 EJ in 2000.

Higher oil pricers have only minor impact on gas demand on a world scale.

In the less developed gas markets (Africa, Asia, Latin America and the Middle East) the industrial sector and electricity generation are the main market sectors accessible to gas on a large scale.

Due to a reasonable increase during the last few years gas demand for electricity generation is now larger than gas demand for the residential and commercial sectors.

There is a continued trend in the gas producing countries towards developing their gas reserves and using the gas in order to decrease oil imports or to increase oil exports.

19

Table 1.8. IGU World Gas Supply (EJ)

	Actual 1986	Potential			
		1990	2000	2010	2020
Moderate Scenario					
Conventional	65.0	75.7	89.6	91.7	92.1
Unconventional	0.5	0.7	2.7	5.7	8.7
Total	65.5	76.4	92.3	97.4	100.8
Rapid Scenario					
Conventional	65.0	78.3	96.9	104.0	108.3
Unconventional	0.5	1.1	4.1	8.8	12.9
Total	65.5	79.4	101.0	112.8	121.2

World Gas Supply Potential

The potential supply of gas in the world under the two IGU scenarios can increase from the 1986 level of 65.5 EJ to a 2000 level of some 92-101 EJ. During the period up to 2020 the supply potential will further increase to some 101-121 EJ, but the annual growth rates will be less than during the previous period (Table 1.8).

Table 1.9. IGU Conventional Gas Supply by Region (EJ)
            (Moderate Scenario)

	Actual 1986	Potential			
		1990	2000	2010	2020
Africa	2.0	3.0	4.1	5.1	6.0
Asia	4.0	5.6	7.5	8.6	9.0
East Europe	26.3	31.8	39.1	39.9	39.9
JANZ	0.7	1.1	1.4	1.3	1.1
Latin America	2.7	3.1	4.3	5.2	5.6
Middle East	2.8	3.1	5.5	6.8	8.3
North America	20.4	21.9	20.4	18.4	16.7
West Europe	6.1	6.1	7.3	6.4	5.5
World Total	65.0	75.7	89.6	91.7	92.1

Potential Production of Conventional Gas. In the case of the moderate scenario, annual worldwide potential production of conventional gas in 2000 is estimated to be about 25 EJ higher (+38%) than the present production. Except for North America (in particular the US) the potential production may increase in each region. In absolute terms East Europe, Asia, the Middle East and Africa will account for the major part of the growth (Table 1.9).

During the remaining years of this century proved recoverable reserves in all regions are higher than the cumulated total production. The remaining reserves in 2000 are sufficient for another 112 years of production at the year 2000 level.

After 2000 the worldwide potential production of conventional gas will continue to rise. The development varies considerably, however, from one region to another. Whereas certain regions show a further increase in potential production (the Middle East, Africa and Asia) other regions (North America, West Europe and JANZ) show a substantial decrease in production. The remaining reserves in 2020 are sufficient on a global scale for another 90 years of production. In all regions, with the exception of North America and Asia, cumulated production until the year 2020 will still be lower than the present proved recoverable reserves.

In the case of the rapid scenario (which is used as a kind of sensitivity analysis) the higher oil price level acts as an incentive, increasing the potential production above the moderate case level, especially after the turn of the century. The difference for the potential production in 2000 is 8% and in 2020 18%.

Table 1.10. IGU Unconventional Gas Supply by Region (EJ)

	Actual	Potential					
	1986	2000		2010		2020	
		m	r	m	r	m	r
East Europe	0.1	0.1	0.1	0.5	0.5	1.3	1.3
North America	0.4	2.6	4.0	5.1	8.0	7.3	11.2
Other Regions	--	--		0.1	0.3	0.1	0.4
World Total	0.5	2.7	4.1	5.7	8.8	8.7	12.9

m = moderate scenario.
r = rapid scenario.

Potential Production of Unconventional Gas. The potential production of unconventional gas is estimated to increase from a 1986 level of 0.5 EJ to 2.7-4.1 EJ in 2000 and to a 8.7-12.9 EJ in 2020 (Table 1.10). This implies that the share of unconventional gas supplies will be less than 3-4% in 2000 and 6-11% in 2020.

The supply potential in 2020 is mainly concentrated in two regions: North America accounts for some 85% of the potential (gas from tight formations) and East Europe for some 10 to 15% (coal gasification).

Compared with the previous studies the trend towards lower estimates for unconventional supplies has been continued. Oil price levels assumed for the present study do not allow an accelerated development of unconventional supplies.

In the two scenarios defined for the study, gas will essentially be produced from conventional sources as natural gas reserves are adequate in nearly all regions of the world. North America is the only region where major unconventional sources are likely to be tapped to maintain potential gas production.

Potential Gas Demand

The base case for the IGU demand outlook is the moderate scenario. The estimates of potential gas demand for each region in the moderate scenario are set out in Table 1-11.

Table 1.11. IGU Gas Demand by Region (EJ)

	Actual 1986	Potential			
		1990	2000	2010	2020
Africa	0.8	1.1	2.1	2.6	3.5
Asia	2.4	3.7	5.6	5.7	6.0
East Europe	24.8	29.9	37.1	38.3	39.1
JANZ	2.1	2.4	2.9	3.5	3.9
Latin America	2.8	3.3	4.4	5.8	7.0
Middle East	1.8	2.3	4.0	4.8	5.6
North America	19.8	21.6	23.3	24.2	24.8
West Europe	8.2	9.0	10.0	10.4	10.6
World Total	**62.7**	**73.3**	**89.4**	**95.3**	**100.5**

Table 1.12. IGU Average Annual Growth Rates of Gas Demand by Region (%)

	Actual 1980/ 1986	Potential			
		1986/ 1990	1990/ 2000	2000/ 2010	2010/ 2020
Africa	17.8	8.3	6.7	2.2	3.0
Asia	10.8	11.4	4.2	0.2	0.5
East Europe	7.0	4.8	2.2	0.3	0.2
JANZ	7.0	3.4	1.9	1.9	1.1
Latin America	2.6	4.2	3.1	2.7	2.0
Middle East	12.2	6.3	5.7	1.8	1.6
North America	-2.5	2.2	0.8	0.4	0.2
West Europe	0.4	2.4	1.1	0.4	0.2
World Total	2.6	4.0	2.0	0.6	0.5

Table 1.13. IGU Shares of Regions in the World Gas Demand (%)

	Actual 1986	Potential			
		1990	2000	2010	2020
Africa	1.3	1.5	2.3	2.7	3.5
Asia	3.8	5.0	6.3	6.0	6.0
East Europe	39.5	40.8	41.5	40.2	38.9
JANZ	3.3	3.3	3.2	3.7	3.9
Latin America	4.5	4.5	4.9	6.1	6.9
Middle East	2.9	3.1	4.5	5.0	5.6
North America	31.6	29.5	26.1	25.4	24.7
West Europe	13.1	12.3	11.2	10.9	10.5
World Total	100	100	100	100	100

Table 1.14. IGU Impact of Higher Oil Prices on World Gas Demand (EJ).

	Actual 1986	Potential 1990	2000	2010	2020
Moderate Scenario	62.7	73.3	89.4	95.3	100.5
Rapid Scenario	62.7	73.1	88.8	94.2	98.8

The table shows that potential world gas demand will increase to about 73 EJ in 1990, to almost 90 EJ in 2000 and to just over 100 EJ in the year 2020. Compared with the actual gas demand of about 63 EJ in 1986 these figures represent average annual growth rates (for total world gas demand) of 2.5% up to 2000 and about 0.5% thereafter.

The main regions as regards gas demand are East Europe, North America and West Europe, which jointly account for about 85% of the gas demand. In these regions the potential annual growth is smaller, however, than the annual growth of gas demand worldwide. Consequently the share of these regions in the total gas demand is expected to decline. On the other hand, the growth in the comparatively small markets is expected to be above average.

If the gas demand potentials, as they are outlined above, were to be achieved as actual gas demand, it is estimated that on a world scale the share of gas in primary energy demand (20% in 1986) will remain within the range of 19 to 21%.

In the mature gas markets (East Europe, North America, West Europe) a stabilization or slight decrease of the market share in the long term is estimated. In the less developed gas markets, however, a further increase of the market share is estimated.

Impact of Higher Oil Prices on Gas Demand. In the rapid scenario the potential gas demand will tend to be smaller. Although the impact varies from country to country, it is estimated that the impact will be relatively small on a world scale.

Market Sector Demand Analysis. A distinctive feature of this study is the analysis of potential gas demand by market sectors in each region. Table 1.15 shows estimates of actual and potential gas demand broken down into residential/commercial, industrial (including chemical feedstocks), electricity generation and other uses.

Table 1.15. IGU Gas Demand by Market Sectors (EJ).

	Actual			Potential			
	1980	1983	1986	1990	2000	2010	2020
Residential/ Commercial	14.9	15.5	16.5	18.2	20.1	21.3	22.4
Industrial	25.4	24.5	23.2	28.5	37.9	41.1	43.8
Electricity Generation	9.0	10.3	17.2	19.9	23.3	24.6	26.1
Other Uses	3.8	4.9	5.8	6.7	8.1	8.3	8.2
World Total	53.9	55.2	62.7	73.3	89.4	95.3	100.5

In absolute terms the highest growth potential is found in the industrial sector, followed by the electricity generation. The share of industrial gas demand (presently 37%), which has been declining over the last few years, is estimated to recover and to reach a level of about 44% again in 2020. The share of the residential/commercial sector will further decline (presently 27%) to a level of 22%. Remarkable is the strong increase of the share of electricity generation during the last years: from 18% in 1980 to 27% at present. A further increase is not, however, considered very likely. The share of the other uses has been remarkably constant over the last years (9%) and is estimated to remain so the forthcoming years.

Area Results of the IGU Study

Africa. African gas reserves are substantial (16,060 BCM/586 EJ) compared with the actual production; the proved recoverable reserves will be sufficient to meet the production estimates until 2020.

The gas supply potential is estimated to increase from an actual level of 2.0 EJ to 4.1-4.6 EJ in 2000 and to 6.0-7.0 EJ in 2020. It is, however, doubtful whether this supply potential will be actually developed, since an important part of the production is intended for exports. The economic viability of these exports is highly dependent on the energy (oil) price development. Recent oil price developments do not warrant commercial development of existing reserves.

Potential gas demand is estimated to increase from 0.8 EJ in 1986 to 2.1 EJ in 2000 and to 3.5 EJ in 2020. These estimates imply annual growth rates of almost 8% until 2000 and of 3% between 2000 and 2020. To realize these high growth figures, major organizational and financial challenges must be met.

Asia. Total gas reserves are assessed at 19,000 BCM (694 EJ), of which 6,400 BCM are proved recoverable and 12,600 BCM additional recoverable. The gas supply potential (actual 4.0 EJ) is not limited by reserves but by market opportunities and by the capability (political, financial, infrastructural) of the countries to develop their reserves.

In the moderate scenario gas supply is estimated to increase to 7.5 EJ in 2000 and 9.0 EJ in 2020. Higher oil prices are an incentive which can lead to a higher supply in the long term (11.5 EJ in 2020).

The rapid growth of the population and the (comparatively) strong economic growth result in an annual primary energy consumption growth rate of almost 3%. Although the market share of gas remains low as compared to other regions (less than 10%) potential gas demand is estimated to increase from 2.4 EJ in 1986 to 5.6 EJ in 2000 and to 6.0 EJ in 2020.

East Europe. Since East Europe largely features centrally planned economies the development of the gas market is not so much determined by market forces (for instance: world energy prices) but rather by the energy planning. On the supply side, as well as on the demand side, the role of the U.S.S.R. is overwhelming. First of all almost 99% of the total reserves are to be found on Russian territory (163,180 BCM/5,956 EJ or some 50% of the total world gas reserves). It is furthermore estimated that during the years to come the USSR will supply more than 90% of the gas in East Europe (1986: 26.4 EJ; 2000: 39.2 EJ; 2020: 41.2 EJ).

After the turn of the century it is expected that some in situ coal gasification or lignite gasification projects will come into production resulting in unconventional gas production of 1.3 EJ (36 BCM) in 2020. The emergence of the gasification of coal is caused by the gigantic and easily accessible coal reserves rather than by the scarcity of natural gas. The additional recoverable gas reserves are sufficient to continue gas supplies after 2020 for more than 100 years, at a level exceeding 36.5 EJ (1,000 BCM) per annum.

Gas demand is estimated to increase to 37.1 EJ in 2000 and to 39.1 EJ in 2020 (1986: 24.8 EJ). More than 85% of this gas demand potential will be used by the U.S.S.R. The high growth rates of the previous years (+7% p.a.) are expected to slow down to a lower level (+3 to 4% p.a.) until 1995. Afterwards growth will probably level off.

JANZ. Total gas reserves in Japan, Australia and New Zealand (JANZ) are assessed at 2,960 BCM (108 EJ). Present estimates of supply are 1.4 EJ in 2000 and 1.2-2.0 EJ in 2020.

The potential gas demand is estimated to increase from an actual level of 2.1 EJ to 2.9 EJ in 2000 and to 3.9 EJ in 2020. This implies an increasing share of gas in the primary energy demand from 11% in 1986 to 13% in 2020.

It is remarkable that presently more than 60% of the total gas demand is used for electricity generation.

Latin America. Total gas reserves in Latin America are assessed at 14,600 BCM (533 EJ). The potential gas supply is estimated to increase from 2.7 EJ in 1985 to 4.3-6.5 EJ in 2000 and 5.6-7.8 EJ in 2020. the highest estimates apply to the rapid scenario.

In general it can be concluded that the limited financial resources of the countries in this region impede the development of gas resources. Gas demand in Latin America is expected to grow from 2.8 EJ in 1986 to 4.4 EJ in 2000 and 7.0 EJ in 2020. These figures imply a market share of gas in the primary energy demand of 17-18%, which is a few percent less than the world market share. Most of the gas is used in the industrial sector (50%).

Middle East. Gas reserves in this region (48,000 BCM/1,752 EJ of which 55% proved recoverable) are the second largest in the world.

The potential supply is relatively low compared to these large reserves: 2.8 EJ presently and increasing to 5.5-6.4 EJ in 2000 and to 8.3-11.9 EJ in 2020. This supply increase reflects the general policy to produce more gas, in order to substitute for oil in the local economies and to increase oil exports. Therefore world oil demand and world oil prices have a great impact on the future development of the gas industries in this region. This can be seen from the differences between the estimates of the two scenarios for 2020 (the rapid scenario gives a 43% higher estimate).

Gas has presently a share of almost 27% in the primary energy demand and this market share is expected to increase to 35%, making the Middle East the region with the highest market share of gas worldwide. The potential gas demand (in 1986 1.8 EJ) is estimated to increase annually by an average rate of 6% to 4.0 EJ in 2000. Afterwards the growth will be lower resulting in a demand of 5.6 EJ in 2020. By that year 50% of the gas consumed will be used for electricity generation.

North America. The North American gas industry has undergone significant structural and economic changes since 1985. During this transition period decreased regional demand for gas has created a large surplus production capacity. This surplus and the growing deregulation of

27

gas supplies and transmission services has led to widespread gas-to-gas competition which in turn has caused gas prices to drop sharply throughout the region.

Gas reserves in North America have declined in recent years by 10% to 38,420 BCM (1,402 EJ), mainly due to lower additional recoverable reserves in the US. The regional gas supply potential however continues to be large. It is estimated to increase under the moderate scenario from 20.8 EJ in 1986 to 23.0 EJ in 2000 and to 24.0 EJ in 2020, maintaining North America as the second largest gas producing region in the world. [The author considers these figures to be quite pessimistic, ie., too low.]

Because of the historically large volume use of gas in North America, unconventional supplies have been the focus of extensive development. In the time period of this study unconventional supplies are expected to make a growing contribution to the region's gas supply. In the moderate scenario for example, in 2000 11% (2.6 EJ) of the total supply potential is expected to come from unconventional sources. By 2020 the share of unconventional supplies is even estimated at 30% (7.3 EJ). In the rapid scenario unconventional supplies are expected to be developed somewhat more rapidly, reaching one third of total supply potential by 2020.

Conventional supply estimates are based on vigorous development of gas resources, with cumulative production from 1986 to 2000 exceeding the current proved recoverable reserves.

Consistent with the steady growth in estimated gas supplies, the North American gas demand is estimated to rebound and grow steadily through the time period of this study. Potential gas demand is estimated to increase from an actual level of 19.8 EJ to 23.3 EJ in 2000 and to 24.8 EJ in 2020. In the rapid scenario the estimates are slightly lower in the long term as compared to the moderate case (-5% in 2020). [Again, these figures are, in the author's view, unduly pessimistic.]

For the time period of this study the commercial, industrial and electricity generation sectors are likely to be the most important growth sectors. A significant portion of electricity generation demand growth is in cogeneration applications, and is therefore reflected in the figures for commercial and industrial demand.

On balance, the North American gas industry is expected to emerge from the current period of regulatory change as a largely deregulated industry with market sensitive pricing. These changes should re-establish the positive growth patterns in both gas supply and demand that characterized the pre-oil embargo years for the industry, and provide good prospects for natural gas as a fuel in North American markets well beyond the study horizon.

West Europe. The total gas reserves in West Europe are assessed at 10,930 BCM (399 EJ), of which 6,730 BCM are proved recoverable and 4,200 bcm additional recoverable.

The supply potential is estimated to increase from 6.1 EJ in 1986 to 7.3-7.9 EJ in 2000 and will thereafter decline to 5.5-6.1 EJ in 2020.

Potential gas demand in West Europe is estimated to increase from an actual level of 8.2 EJ to 10.0 EJ in 2000 and to 10.6 EJ in 2020. The market share of gas will further increase to 17% in 2000, but subsequently there will be a decrease to 15% in 2020.

REFERENCES AND NOTES

1. International Gas Union, *Report of Committee J; World Gas Supply and Demand, 1986-2020*, Paris, France, June 1988, pp. 4-5.

2. Ibid., pp. 33-34.

3. Ibid., p. 35.

4. American Gas Association, *A Comparison of Remaining World Gas and Oil Resources*, Arlington, Va., December 17, 1982.

5. British Petroleum, *BP Review of World Gas*, London, England, July 1988.

6. Institute for Energy Resource Studies, Colorado School of Mines, *A Discussion of Resource and Reserve Definitions for Petroleum, Natural Gas, Coal, Oil Shale, Uranium, and Ore Deposits*, Golden, Colo., March 1987.

7. Gustafson, T., *The Soviet Gas Campaign*, Santa Monica, Calif., Rand Corporation, p. 31, June 1983.

8. "1989 Outlook," *Gas Energy Review*, Vol. 17, No. 1, January 1989.

9. "Northeast Europe Pursues Diverse Agenda," *World Oil*, April 1989.

10. U.S. Department of Energy, Energy Information Administration, *International Energy Annual*, Washington, D.C., 1980-1986.

11. American Gas Association, *LNG Fact Book*, Arlington, Va., December 1987.

ACKNOWLEDGMENTS

The author/editor wishes to express his thanks to the authors and copyright holders for permission to draw heavily upon the works incorporated into this chapter. Significant portions of the material were contributed by Robert Kalisch, Paul McArdle and Michael German. Copyrighted materials were quoted with permission from the American Gas Association and the International Gas Union. The draft was typed and proofread by Lisa Hamako.

2

The Economics of Natural Gas Development

Afsaneh Mashayekhi

Some of the significant economic issues that energy policy makers face in planning natural gas development and consumption are covered in this chapter. This discussion is limited to economic cost estimating and pricing of natural gas, as well as a practical methodology for gas utilization.

Natural gas is one of several major energy resources, and the fundamental economic principles that govern the appraisal of any natural resource project apply to natural gas. The importance of these economic principles is that they help energy policy makers assess the cost and value of natural gas to the economy. Only when the economic appraisal of a project indicates that the value of developing gas is higher than alternative sources of energy should the project be considered for implementation. Related costs and benefits of alternative fuels should also be included in such an appraisal.

The key issues behind gas development are economic issues. Ignorance about the economics of gas has slowed down gas development in many countries. One obstacle to gas development has been the structure of gas costs and pricing. There is no single cost and price estimation standard that makes comparison with alternative fuel costs easy. Another obstacle to the development of gas has been the determination of the high value uses of gas. In many countries gas has been treated as a "premium" fuel that should not be used as a boiler fuel even when gas costs are quite low and gas reserves are plentiful.

This chapter will address the issues of estimation of gas costs and pricing of natural gas in such a way as to provide sufficient incentives for producers and a fair "rent" to the government, as well as the economic value of gas in different uses.[1]

Estimation of Marginal Costs

Economic theory holds that, in order to provide an efficient allocation of resources, the price of a good should be equal to the marginal cost of expanding the output of the good so that it clears the market. Marginal cost pricing theory dates back to the works of Dupuit and Hotelling.[2] In the 1950s, Boiteux and others developed the theory for application to electric power in particular.[3] In the 1970s, more sophisticated models were developed to estimate marginal costs, but there has been less work on applying marginal cost theory to natural gas.[4]

The rationale for setting prices equal to marginal cost is that net benefit is maximized at the point when price equals marginal cost, and demand and supply curves intersect. Marginal cost pricing theory ensures that the customer signals the justification for further investment by his willingness to pay the marginal cost incurred. This is contrary to the traditional practice in many gas utilities, where gas costs are defined as embedded or accounting costs.

The accounting cost approach implies that resources to be used in the future cost the same as those used in the past. This could lead to over- or under-investment depending on whether resource costs are actually increasing or decreasing. Similarly, the average cost method would not reflect the cost components with large economies of scale.

The average incremental cost (AIC) approach is one of the more satisfactory and widely used methods for calculating the long-run marginal cost. It is estimated as the ratio of the discounted incremental costs of meeting future consumption by the discounted volume of incremental supply.[5]

$$AIC = \frac{[\,I + (R_t - R_o)\,]\,/\,(1 + i)t}{(Q_t - Q_o)\,/\,(1 + i)t}$$

i = discount rate

I = marginal capital cost

$R_t - R_o$ = marginal operating and maintenance costs

$Q_t - Q_o$ = marginal supply

Under ideal conditions, in a perfectly competitive market, with perfect information and no costs associated with adjusting prices, short-run marginal costs should be estimated. In practice, the limitation imposed by normal economic circumstances such as externalities, sub-optimal system plans, imperfect information and uncertainty make departures from "first-best" solutions necessary.[6] Also, it may be difficult politically and administratively to have frequent large fluctuations in prices each time a lumpy new investment is made. Therefore the long-run average incremental approach is a more practical and acceptable method of cost estimation.

In estimating costs it is also important to distinguish between peak and off-peak costs. In a single model, off-peak consumers should only be charged the operating and maintenance costs, since the system has spare capacity during those periods and pressure on capacity is entirely due to peak consumers. Therefore, the economic cost of gas should have two components corresponding to peak and off-peak demand.

There are three broad categories into which marginal costs are frequently allocated: energy costs, capacity (or demand) costs, and customer costs. Marginal energy costs cover the capital and operating cost of gas exploration and development, and vary in direct proportion to the value of the gas. Marginal capacity costs are basically the costs of facilities needed to meet an increase in demand. Included in these costs are: marginal capital and operating costs for transmission and storage and the portion of production and distribution costs related to gas used at peaking time. Customer costs are directly related to the number of customers served and include the costs of connection, metering and billing. They are estimated as the costs of a hypothetical skeleton distribution network to serve minimum-sized load (such as minimum-sized distribution mains services), the meters necessary to connect customers, operating expenses incident to customers (such as meter reading), and other costs that are a direct function of the size of the customer group. These may be charged to the consumers as a lump-sum connection charge or as a fixed periodic payment.

In practice it is difficult to differentiate clearly between customer-related costs and capacity costs. The use of the hypothetical minimum system, as described above, often results in customer costs that are too high. It is probably preferable to arrive at the capacity costs of distribution directly, rather than by subtracting customer costs from total distribution costs. The capacity costs can be estimated by analyzing the components of the gas distribution system that vary with demand. The following sections discuss the typical components of a gas system and the steps involved in system planning to formulate investment plans and, therefore, system costs.

Gas System Planning

To determine the marginal cost of supplying gas, economists and engineers in gas companies work closely following an iterative process to plan the system. The planning model generally uses linear or dynamic programming methods to estimate demand and load forecasts and develop capacity and supply needs. Its goal is to minimize an objective function: the present value of system costs to meet end-use demand subject to constraints. These constraints include the requirement to meet the load forecast, limited access to storage for peaking, and restrictive supply contracts.

Before an investment plan is made, it is necessary to prepare demand forecasts. Both the level (average volume in, say, thousand cubic feet, MCF) and the structure of gas demand (MCF per hour) are required for gas system planning. The structure of demand includes disaggregation by geographic area and consumer category, as well as local characteristics such as the load factor.

The load factor is the ratio of average to maximum or peak demand over a given interval of time. It is important because the capacity of the system, and consequently its costs, are largely determined by the ability to meet peak demand. The combination of gas sources used is determined by different consumers, fluctuations from hour to hour, day to day, and season to season. The high peak occurs only over a short period of time. The peak demand is often met by stored gas or by interruptible gas sources. The shoulder demand is met by pipeline gas and some storage. During the base load period, demand is met from pipeline gas while surplus production is put into storage. Of course, the peak demand does not occur for all consumer groups simultaneously. The "diversity factor" for a group of consumers measures the divergence of individual peak loads from the system average.

Supply decisions include choosing which sources of gas to use and when to use them during any year. Therefore the planning model develops a plan for dispatch of each gas source to meet seasonal demand patterns. Gas planning needs to integrate the exploration, development and transportation phases. Individual companies, however, break this process into relevant sub-systems.

The Application of
Marginal Costs to Pricing

Once the strict long-run marginal cost is calculated, both its level and its structure should be adjusted within an economic framework to meet second-best and non-economic objectives. Marginal costs must often be adjusted to take into account financial, social, political and other considerations that are important in setting prices. Practical considerations such as difficulties in billing and collection should also be considered at this stage.

For example, where prices elsewhere in the economy do not reflect marginal costs, a second-best departure from a strict marginal cost pricing policy for gas may be required. If petroleum products are taxed, the first-best solution would be to revoke those taxes. But if energy policy or fiscal objectives require fuel taxes, then a second-best price for gas should be set above marginal costs.

One desirable way of adjusting prices is on the basis of the Baumol-Bradford inverse elasticity rule, whereby the greatest (least) divergence from strict marginal cost occurs for the consumer group and pricing period where the price elasticity is lowest (highest). This rule would result in the smallest deviations from the optimal levels of consumption consistent with the strict application of the marginal cost pricing rule.

Further, subsidized or life-line gas prices for residential consumers in certain poorer areas in cities may be justified on equity grounds if gas prices are high relative to income levels. In theory, however, a lump-sum transfer to low-income households to meet these overall basic levels may be preferable to specific subsidies. Also in practice in most countries, gas distribution is concentrated in urban areas inhabited by higher-income groups who pay less for gas than lower-income groups pay for LPG and kerosene. Therefore, although the concept of a life-line rate is appealing, it should be used only in specific cases and not misused to subsidize high-income groups.

It is also necessary for gas prices to ensure cost recovery for gas companies. The principal financial objective in gas companies is generally related to their revenue requirement, such as a target financial rate of return on assets. A widely-used criterion of financial viability is the ratio of a utility's net income after taxes to net fixed (revalued) assets in operation. In many private utilities, regulatory authorities require a "fair" rate of return as an upper limit on earnings.

Another price adjustment may be due to the practical difficulties and costs of metering and billing. An important factor is that the tariff structure should be comprehensible to the average customer. The degree of sophistication of metering should be determined by the net benefit of metering, the practical problems of installation and maintenance and the ease of billing and collection. For example, the amount of gas that residential consumers without space heating can consume is limited, while the cost of metering residential demand is relatively high. Therefore, for residential customers in countries such as Bangladesh, it may suffice to bill on the basis of the number of gas-using appliances they have, rather than to introduce costly metering.

Demand has a seasonal pattern which affects marginal gas costs. During summer or base load months, little gas is used for heating purposes. In winter, when the demand for space heating increases, or at times when hydropower generation falls because of droughts, the demand for gas increases. In countries or regions where demand has a seasonal pattern that affects marginal gas costs, it is important to have different rates for summer and winter reflecting the different marginal costs. The demand and load factor characteristics of various consumer classes can differ significantly. Certain industrial customers have high load factors while some power or residential customers experience large fluctuations during the day or seasonally. Differential rates assure that each class will pay a rate that reflects the marginal costs it imposes on the system over the year.

The average incremental cost methodology has been used to estimate costs in eight developing countries.[7] These countries include a range of reserve sizes, distances to the market, and location (off-shore and on-shore).[8] The costs of gas exploration and production in this sample range from $0.61 to $1.29 per MCF. Similarly, costs of gas delivered to the city-gate are estimated at $1.10 to $1.80 per MCF. This data is only indicative of costs in these countries; it demonstrates that gas costs are lower than expected in most cases. In fact, even with crude oil prices of $15 per barrel, these gas prices are a fraction of alternative fuel costs. Further, since many gas fields in developing countries are at an initial development period, costs are expected to remain at their low levels.

Obviously, these costs will change over time. Demand for gas will shift over time as gas becomes available. Therefore initially as a result of higher production levels and economies of scale, the marginal cost of gas will fall. Once the most accessible and easily exploitable resources are depleted, the marginal cost of gas may begin to increase. These estimates of natural gas costs therefore need to be used within the context of the dynamic natural gas supply and demand.

The overriding criterion for gas pricing should be economic efficiency, while ensuring that the financial viability of gas companies is upheld. An economically efficient price will clear the market while a higher price will induce producers to supply more than consumers are willing to buy at the higher prices, and vice versa. In addition to clearing the market, efficient gas prices will encourage additional production whenever the expected costs are less than the value of incremental supplies, and on the demand side they will discourage wasteful consumption.

Some of the fundamental issues in pricing natural gas are the concepts and principles which should govern the setting of producer and consumer gas prices, and the setting of pipeline tariffs. There is also a problem of uncertainty of both future international energy prices and future domestic prices of alternative fuels. Early agreement on pricing is an important contribution to the success of a project. The sharing of natural resource rents between the government, consumers and private companies also needs to be determined at an early stage.

This is because a distinctive economic attribute of an exhaustible resource such as gas is that its economic price is composed of its production cost and an additional component, sometimes called "rent" or depletions premium, to reflect the cost of future consumption forgone by using the resource today. The fundamental principle of depletable resource economics is that under equilibrium conditions, the depletion premium must increase over time at a rate equal to the opportunity cost of capital. Thus the principal distinction between a depletable and a renewable resource is that the economic price of the former will include a depletion premium as well as its production cost. A further point needs to made about the application of the Hotelling principle to gas prices. In cases where gas reserves are very large compared with demand, the depletion premium will be a theoretically valid but quantitatively insignificant part of an economic efficiency price. In intermediate cases, it may be large enough to be important, and its estimation will regain estimation of future extraction costs, the demand patterns, the long-run trend of discount rates, the resource size and the future price of the substitute.

One of the thorniest problems in setting prices for depletable resources is that of how the surpluses should be shared between the different economic units: the owner of the resource (often the government), producer companies, and the consumers. If tariffs are set at marginal production costs, all the surplus will accrue to consumers. Producers will be compensated for their costs but no incentives will be provided to prospect for additional resources and to consume sparingly.

The price to the consumer should at least include the marginal cost and the depletion premium to provide the proper incentives for conservation of an exhaustible resource and exploration for additional resources.

Further, in determining gas prices a distinction needs to be made between gas prices paid to producers, gas prices paid by consumers, the opportunity price of gas, the value of gas to the economy. The latter is the value of gas determined by the value of its substitute plus net external benefits less relevant opportunity costs.

The overriding rationale governing pricing is that prices to consumers and producers should reflect the economic cost of gas, including any depletion premium. This means that the consumer price should be equal to the consumer cost. The schedule of consumer prices to different consumer types should reflect their load factor, seasonality of use, and volume. The transfer price between the distribution company and the transmission company should be equal to the consumer price less a margin to cover the marginal cost of distribution and ensure the financial viability of the distribution company. Similarly, the transfer price between the transmission company and producers should be equal to the previous transfer price less the marginal cost of transmission. The producer price should reflect the marginal cost of exploration and production but also divide a negotiated portion of the rent between the producers and the government. Therefore, an efficient pricing system would provide incentives to explore and produce as well as to consume efficiently while returning some rent to the government.

Comparison of Gas Use Alternatives

For a number of developing countries, indigenous natural gas resources hold the key to reducing expensive reliance on oil, as well as providing a cleaner fuel source. Yet in many cases, the current level of gas use is still low relative to the reserve base and to potential demand. Embarking upon the development of natural gas -- a non-renewable fuel and one which requires large, up-front investments to transport and use -- raises complex questions of gas allocation and investment strategy that must be faced at the pre-investment stage of development. Many of these questions, as shown below, are similar across countries.

Should gas be used in electricity generation to replace imported fuel oil or coal, or should a gas-based fertilizer plant be built to replace imported urea?

Would the high cost of a city gas distribution network be justified by the very high and growing cost of the kerosene and LPG used by households that gas could replace?

If gas reserves are large, should the country try to attract commercial partners for an LPG export project or would it be better to keep the gas in the ground to satisfy future growing domestic markets?

As a producer of fertilizers or petrochemicals for export, could the country compete with supplies from the Middle East or Mexico where production is based on associated gas?

There are two things to note about such questions. First, they are essentially economic rather than technical questions. The technical feasibility of using gas for power generation, for fertilizer production, for LNG, and so forth, has been well-proven in other countries and need not be established anew. The important issue is the relative economic merit of the different alternatives. This will depend upon such country-specific parameters as the amount of base-load hydro in the power system, the proximity to major export/import markets for urea, and the density of housing in urban areas. In many cases the economic value of gas in thermal power generation (displacing diesel fuel oil or wood) and in industrial applications exceeds the value in LNG exports by a wide margin.

The second point to note is that although such questions concern specific project alternatives, they are really about long-run sectoral strategy. What should be the role of gas in a country with abundant lignite or hydro resources? Can gas be used as an engine of economic growth through its "embodied export" as urea or LNG? Through which sectors can the penetration of gas as an oil import substitute be most rapidly achieved? How could such substitution be phased out later if supplies become scarce and can be allocated only to the highest value uses?

Because gas is depletable, such trade-offs over time, as well as the trade-offs between gas-using projects at any point in time, must be explicitly considered. Further, because gas infrastructure is lumpy, it should generally be designed on the basis of total, long-run gas demand rather than that of an individual project. For these reasons, a long-run, sector-wide framework is generally necessary for the economic evaluation of gas-using projects and for the design of gas investments. The development of such a framework is the first objective of a gas utilization study.

Following from that sector-wide framework, the second objective of such study, is to provide an early comparative screening of potential gas-using and gas-producing projects. It should also define the critical but currently uncertain parameters that affect individual projects and provide direction for the appropriate focus of the next round of feasibility or design studies.

It should not be expected to make final project selection decisions for the country's gas development program over, say, the next two decades. That would be impossible, given the large and numerous uncertainties inherent in pre-investment analysis. Instead the study should be regarded as a preliminary but rigorous planning exercise that produces specific recommendations for the next steps in gas development and that becomes the basic management tool for future strategic analysis, as more information and new options arise in the gas sector.

While the theoretical framework and the practical methodology discussed in the preceding sections have been designed to cover all cases of gas utilization studies, it is clear that the critical issues of gas use in a gas short country such as Pakistan will be much different from those of a gas surplus country such as Bangladesh. Different aspects of the analysis will assume primary importance, and other aspects will be incidental in each case. Although this section focuses on country-wide markets, it is also possible to have both gas short and gas surplus situations in the same country when the market is segmented. We will consider the application of the methodology to three types of situations: gas surplus, gas short, and surplus window.

Gas Surplus Countries

A gas surplus country can be defined as one in which the demand/supply balance is such that the point of economic depletion is very far into the future. If this point is 40 or 50 years away, the present value of the depletion premium becomes insignificant, compared with the range of uncertainty surrounding other estimates such as the marginal gas costs. Thus, for all practical purposes, the economic price path for gas over the long-run becomes the long-run marginal cost of gas development. This implies that for a gas surplus country relatively little time should be spent on developing estimates of gas supply or aggregate gas demand (beyond what is necessary to establish that the situation is indeed one of gas surplus), and relatively more time defining the short- and medium-term investment program in gas production and infrastructure to provide a firm basis for the marginal cost estimates. Alternative gas reserve and supply scenarios are probably unnecessary, even where there is considerable reserve uncertainty, because it will make little difference to the analysis whether the point of depletion is 50 or 150 years away. Similarly, aggregate demand projections can be fairly general and focus on the long-run.

At a later stage, a gas surplus country should try to assemble a base case package of domestic gas using projects that take full advantage of the opportunities for gas penetration in the country's energy sector. Questions of project sequence will not be paramount, since gas availability is not a constraint, so that increases in net present values will come mostly from bringing projects onstream more rapidly. The constraints to this will often be economy-wide: availability of capital, managerial skills, industrial infrastructure, etc. Therefore, careful attention should be paid to exploiting potential complementarities and economies of scale in gas investment. Project packages should be designed and costed to include expatriate supervision or construction where that can be effectively used to overcome implementation constraints. When considering large export-oriented projects, it will often be helpful to initially treat each candidate as mutually exclusive. This will provide a comparative screening of the export options to see which are the most attractive ones, before packages involving the export projects are formulated.

The sensitivity analysis should include the often major technical, financial and market risks associated with any large and/or export-oriented projects that are included in the selected packages. This may require a regional analysis of the market for, say, urea in the surrounding countries. Because the returns on such capital-intensive projects are often highly sensitive to the rate of capacity utilization, it is important to establish the boundaries of such parameters that would be required for the project to be economically desirable. This type of broad brush analysis should be undertaken in the gas utilization study rather than left to further feasibility work on each export project alternative. It is both possible and desirable to use the sensitivity analysis portion of a gas utilization study to screen out those export-oriented projects that have less chance of success than others, and thereby to focus future feasibility studies on those candidates with the greatest potential.

Gas Short Countries

A gas short country is one in which the potential gas availability is projected never (or only briefly) to exceed the potential demand. In a number of countries, current fuel oil consumption is several times as large as natural gas production. Even if the latter increases significantly, with projections of continued economic growth it is likely that incremental gas supplies will continue to replace fuel oil at the margin. In such a case, the economic price path for gas will follow that of fuel oil, so long as the cost of gas development remains below that level. Since consumption will be supply-constrained in a gas short country, in a gas utilization study more time should be spent on the supply side than on the demand side. It will often be important to develop alternative supply scenarios and to plot the maximum supply path for each. The costing of gas development can be done roughly, simply to ensure that costs remain below the cost of the replacement fuel. Demand analysis should focus on identifying specific project candidates in the main domestic markets (e.g., electricity fertilizer, cement) rather than estimating long run trends in aggregate demand.

Project packaging, ranking and selection are the most critical parts of a gas utilization study for a gas short country. With gas availability as a constraint over the entire period, questions of project size and sequence are of major importance. With a high gas value, the selection of project candidates is probably limited to those geared towards the domestic market. It will often be useful to pre-screen project candidates for inclusion by calculating their gas netback values and comparing them with the value of gas. Then trial packages can be formulated from those projects passing the netback test, and net project values can be calculated on permutations of each trial package.

The consistency check with aggregate supply may show discrepancies which will require repackaging of projects. The sensitivity analysis will probably focus mostly on the effect of different supply scenarios and their implications for project selection and timing. These could be critical if one of the supply scenarios is optimistic enough to take the country into the surplus window category as described below.

Surplus Window Countries

A surplus window country is one for which a 20-30 year period includes times of both gas surplus and gas shortage. The point of supply catching up with demand is either past or imminent; depletion is either projected within the gas utilization study period or clearly forseeable beyond it. In such a case, there are few short cuts to the full methodology that can be taken. Both demand and supply must be estimated carefully in order to derive the gas price path. A general estimate of the cost of gas development will be needed to ensure that the price path does not fall below it. It may be useful also to pre-screen project candidates by calculating their netbacks and slotting each project into the time period at the point where its netback is first higher than the economic price of gas. Some projects will become viable only once gas supply can satisfy demand and the price of gas falls. These should be compared initially on a mutually exclusive basis to find the best candidates for that possibly brief period of gas surplus. If more than one or two such projects are included in the package for that period, the consistency check may show that the resulting level of aggregate gas demand in the first years of gas surplus will raise the gas price path enough to replace the marginal projects.

Both consistency checks and sensitivity analysis are important to ensure robust results. For a surplus window country, the addition of a single large gas-using project to the base case project package can result in a large enough shift forward in the date of depletion to significantly raise the gas price path during the original projected surplus period. This would require a new calculation of the net project values for the larger project package using the revised price path. A comparison of the new net project values with that of the base case (using the original price path) will explicitly incorporate the trade-offs involved in putting the gas to earlier productive use, versus saving it for higher valued uses in the future. Sensitivity tests should be designed to highlight the physical and economic uncertainties to which the results are most sensitive.

REFERENCES AND NOTES

1. For a more detailed overview see *The Economics of Natural Gas: Pricing, Planning and Policy*, Chapter 4, Julius, De Anne and Mashayekhi, Afsaneh, Oxford, England, Oxford University Press (Forthcoming).

2. Dupuit, P., "De l'utilite et de sa mesure," La Reforma Sociale, Turin 1932, and Hotelling, "The General Welfare in Relation to Problem of Railway and Utility Rates," *Econometrica*, Vol. 6, July 1988, pp. 242-269.

3. Boiteux, M., "La tarification des demandes en pointe," *Revue Generale de l'Electricite*, vol. 58, 1949.

4. For applications to natural gas see Mac Avoy, Pindyck, *The Supply of Natural Gas Resources*, 1980.

5. See Julius and Mashayekhi, op. cit.

6. Baumol, W.J. and Bradford, D.F., "Optimal Departures from Marginal Cost Pricing," *The American Economic Review*, pp. 265-282.

7. Cameroon, Egypt, India, Morocco, Nigeria, Tanzania, Thailand and Tunisia.

8. See Julius and Mashayekhi, op. cit.

3

Natural Gas and the Environment

Nelson E. Hay

PART 1: OVERVIEW

In the 1970's natural gas was written off in the thinking of many energy planners. It was a "buggy-whip" industry. Today, it is being called "the fuel of the future" -- at least for the next few decades. Why has this turnaround occurred?

As discussed in Chapter 1 and elsewhere in this book, there have been several important factors. Improved gas exploration and production technology has been one, and with it an expanded perception of the natural gas resource base. Similarly, improved gas utilization technology has greatly increased the efficiency of gas use, effectively increasing the life of the gas resource and rendering gas applications more competitive in the marketplace. While this has been true in all sectors -- residential, commercial, industrial and vehicular -- developments in gas turbine technology have particularly revolutionized the economics of electricity generation.

Related to this, the problems which other energy forms have encountered have also been an important factor encouraging a re-evaluation of the role of gas. For example, in the 1970's, large-scale coal and nuclear electric powerplants were generally regarded as the economically preferred means of generating electricity. This situation was expected to continue until solar and other renewable energy forms became viable before the end of this century, with fusion to follow.

The reality has been that rapidly escalating costs of coal and nuclear energy, as well as associated environmental and regulatory barriers have diminished the attractiveness of these options. In the United States it is unlikely that a new nuclear powerplant can be ordered, sited, constructed and brought on line before 2010. Incidents such as the Three Mile Island accident and more recent public revelations regarding US government nuclear facilities have greatly increased US public resistance to nuclear plant siting and waste disposal. Indeed, one nuclear plant under construction in the US has been converted to natural gas, and there are proposals to convert several more, including plants already in nuclear operation. International nuclear incidents such as Chernobyl, and other

47

recent revelations regarding accidents at Soviet nuclear facilities, combined with greatly increased worldwide communications, appear to be decreasing public acceptance of nuclear power on a world scale, as has already occurred in the US. Increased use of coal is similarly being discouraged by increasing perceptions regarding the role of coal in exacerbating such international environmental problems as acid rain and climate change (the "greenhouse effect").

The promise of renewable energy and fusion has not fared much better. With both oil and natural gas in excess supply during much of the past decade, research and development activities regarding these options have waned. In spite of occasional flurries of excitement, a significant contribution from economically viable renewable and fusion or other new energy sources seems unlikely (though the unexpected can happen tomorrow) before 2030-2050.

On the other hand, it seems quite clear that our growing environmental problems -- and growing awareness of those problems -- will drive us toward significant innovation in our energy and resource technologies in the interim. As stated in Chapter 1, protecting natural resources -- as well as public health -- is increasingly being seen at the developing country level as a necessary survival strategy rather than a luxury, a means of sustaining progress rather than constraining it. With the great increase in population projected, combined with widening expectations of economic development, environmental pressure will in all likelihood increase dramatically over coming decades.

It is difficult to accurately project the societal and technological responses to this pressure. Resource conservation and reduced air and water impacts will almost certainly be priorities. Reliance on fossil fuels may ultimately decline, or we may find altogether new ways of finding, producing and using those resources with dramatically reduced environmental consequences.

Which brings us to another of the factors behind the resurgence of natural gas, one that is probably the single most important driving force behind future natural gas technology and use: the environmental benefits of gas. Whatever environmental/technological course the long-term holds, increased use of natural gas in the interim is a directionally-correct strategy. Natural gas is an abundant resource which may itself prove to be important far beyond 2050. Or, it may be technologically displaced. In either case, increased natural gas use in lieu of other fossil fuels and biomass can significantly reduce economic and environmental costs in the interim.

In many nations, increased reliance upon natural gas resources in national energy planning offers opportunities to simultaneously minimize pollution, including the build-up of "greenhouse gases," while also choosing a least-cost energy strategy. Natural gas resources are abundant, widely distributed, and highly underutilized. By all measures -- air emissions, solid wastes, water pollution, and water consumption -- natural gas is by far the most environmentally desirable fossil fuel. Compared to other energy alternatives, natural gas systems are also highly capital and resource efficient, requiring significantly lower capital investment to produce, deliver and utilize a unit of energy. Recently commercialized utilization technologies have greatly enhanced these advantages.

In electricity generation, natural gas is emerging as the international fuel of choice. Natural gas-fueled electric generation offers lower capital and operating costs, much shorter construction lead times, high efficiency, rapid start-up, increased reliability and minimum environmental impact. Gas-fired electric-generating equipment also provides planning flexibility by easily accommodating incremental capacity additions that can track electric load growth more closely. In many applications, natural gas can complement coal use by increasing combustion efficiency and significantly reducing emissions.

Environmental Impacts of Natural Gas

Environmental impacts are of concern both at the point of end-use energy consumption and at all points in the chain ("full cycle") of production, processing and distribution. The environmental advantages of natural gas relative to other energy forms are equally impressive from either perspective.

Full-Cycle Emissions

For example, as discussed later in this chapter, a study of pollution associated with residential sector energy consumption in the United States found that when emissions from the full energy cycle are considered, use of natural gas space heating and other appliances results in only 15 to 20 percent of the total air emissions, and less than one percent of both the total water pollutants and noncombustible solid wastes compared with equivalent use of electric appliances.

Natural Gas Combustion

From the end-use or power plant combustion perspective, natural gas is also inherently cleaner than the combustion of other fossil fuels. (See Table 3.1). For example:

Sulfur Dioxide. High concentrations of sulfur dioxide (SO_2) in the atmosphere can adversely affect health by irritating the eyes and upper

Table 3.1. Typical Annual Emissions from Large New Stationary Sources with Pollution Controls by Fuel Type (Tons/Year)

	Sulfur Dioxide	Nitrogen Oxides	Carbon Dioxide	Total Hydro-carbons	Particu-late Matter
Large Industrial Boiler[1]					
Natural Gas	0.4	74	42,815	1	4
Resid. Oil-Scrubbed	120	149	63,290	4	37
Coal With Scrubber					
High Sulfur[3]	212	260	75,205	2	19
Low Sulfur[4]	61	260	75,205	2	19
240 Megawatt Electric Powerplant[2]					
Gas Combined Cycle	3	1,352	622,092	16	9
Resid. Oil Scrubber	2,200	1,855	1,168,240	69	206
Coal With Scrubber					
High Sulfur[3]	4,123	2,385	1,388,144	34	206
Low Sulfur[4]	3,436	2,385	1,388,144	34	206
Uncontrolled					
High Sulfur Coal	50,050	7,000	1,388,144	34	20,600
Residual Oil	8,500	2,600	1,168,240	69	700

Note: Emissions of sulfur dioxide, nitrogen oxides and particulate matter based on EPA's new source performance standards currently in effect for new large industrial boilers and new electric generating units. No NSPS have been set for total hydrocarbons, and therefore uncontrolled emission factors published by EPA were used. See: *Federal Register*, Vol. 49, No. 119 (June 19, 1984) P. 25106. Based on high efficiency units -- coal and residual oil, 34 percent; combined cycle, 45 percent. Carbon dioxide emission factors, not set out by EPA, based on: "Carbon Dioxide Emissions from Fossil Fuel Combustion . . . ," American Gas Association, (Arlington, Va.: September 1977).

[1] MMBtu per hour, 85 percent capacity utilization.

[2] 65 percent capacity utilization.

[3] 3.5 percent sulfur, with scrubber and electrostatic precipitator.

[4] 1.0 percent sulfur, with scrubber and electrostatic precipitator.

respiratory system, and by possibly reducing air flow through the lungs. In addition, SO_2 can weaken cotton, leather and a variety of other natural fabrics and building materials, such as iron, steel and particularly limestone.[1]

The subject of acid deposition is also closely tied to SO_2 emissions. Sulfur oxides (primarily SO_2) are subject to chemical changes in the atmosphere and return to the earth as acid compounds either in rain or snow, or as a dry deposition. There is a general consensus in the scientific community that acid deposition can cause serious ecological damage.

Sulfur dioxide emissions from natural gas combustion are virtually zero. A boiler operating on natural gas emits, on average, only 0.0006 pounds of SO_2 per million Btu (MMBtu) of fuel burned. In contrast, "typical" grades of fuel oil and coal consumed in large boilers (e.g., 1.5 percent sulfur oil; 2 percent sulfur Eastern coal, uncontrolled) emit 1.6 pounds of SO_2 per MMBtu and 2.6 pounds per MMBtu, respectively.[2,3]

Particulate Matter. The term "particulate matter" refers to small, man-made or natural substances that are light enough to become airborne. An example of man-made particulates is smoke from industrial processes. Natural particulates include dust, pollen and sea salt.

Particulate matter at high concentrations can impair respiratory functions, reduce visibility and discolor the air. Fine particulates are the most damaging to human health. Particulate matter also may affect our climate, as it is believed that high concentrations of particulates reflect solar radiation back into space and change precipitation patterns. Particulate matter can cause damage to many materials used by man. It can discolor paint, weaken fabric, fade textiles, corrode metals, etc. Particulates blown by the wind also can cause erosion of buildings, bridges, and other man-made structures.[4]

Natural gas combustion emits almost no particulate matter. The burning of natural gas results in particulate emissions averaging only 0.01 pounds per MMBtu of fuel consumed. In contrast, the combustion of 1.5 percent sulfur fuel oil emits 0.12 pounds of particulate matter per MMBtu, and coal with a 10 percent ash content emits 7.60 pounds of particulates per MMBtu (uncontrolled).[3] Various control devices are used to limit particulate emissions from large coal-burning facilities. For example, electrostatic precipitators can reduce the particulate emissions at a large (500 megawatt) powerplant by over 98 percent at a capital cost in the range of $50 million.

Carbon Dioxide. Carbon dioxide (CO_2) is the principal cause of the postulated earth warming known as the "greenhouse effect." Natural gas combustion produces less carbon dioxide than coal or oil due to the lower carbon/hydrogen ratio in methane. In addition, in many applications natural gas equipment is more efficient than coal- or oil-fired equipment; fuel consumption and emissions are therefore reduced. In boilers, natural gas emits roughly one-half as much CO_2 per unit of energy output as coal and 30 percent less than oil.[5] In a comparably sized combined-cycle unit, gas emits only about 40 percent as much CO_2 as a coal-electric unit. It has been estimated that in vehicular applications, natural gas use results in about 30 percent less CO_2 (including refinery emissions) emitted than when the same vehicle is fueled with gasoline or diesel.[5]

Nitrous Oxide and Other Nitrogen Oxides. Nitrous oxide (N_2O) is suspected to cause 2-5 percent of the greenhouse effect.[6] The primary anthropogenic (i.e., resulting from human activity) sources of N_2O are believed to be the burning of relatively unrefined fuels with high nitrogen content -- such as coal, residual oil and biomass -- and, use of nitrogenous fertilizers.[7] Specifically, coal combustion is believed to be the dominant anthropogenic source of N_2O.[8] Natural gas combustion results in minimal N_2O emissions.

Nitrous oxide is only one of the nitrogen oxides (NO_x). Nitrogen dioxide (NO_2) and nitric oxide (NO) are of interest for their potential role in formation of ozone, which is a greenhouse gas. About 90 percent of all NO_x emissions are in the form NO which, in the presence of sunlight, ozone and hydrocarbons, forms NO_2. This conversion is the first step in a series of photochemical reactions, which, under certain meteorological conditions, produces smog. Photochemical smog may have a number of adverse effects on human health and on various materials.

In most applications *total* NO_x emissions are 35-90 percent lower with natural gas than when using other fossil fuels. Total nitrogen oxides emissions from large boilers operating on coal may be more than three times as great as emissions from gas-fired boilers. Depending on the coal and boiler type, NO_x emissions from large coal-fired boilers (uncontrolled) range from 0.9 to 1.7 pounds per MMBtu of fuel combusted. In contrast, comparably sized gas-fired units emit only about .14-.55 pound of nitrogen oxides emissions per MMBtu.[3] In a combined-cycle turbine unit, NO_x emissions resulting from gas use are only 10 percent of those resulting from coal use to generate an equal unit of electricity.

In vehicles there is great potential to achieve significant nitrogen oxides emissions reductions by utilizing natural gas in engines designed for natural gas.[9] Already, natural gas buses tested by the US Environmental Protection Agency (EPA) have demonstrated nitrogen oxides emissions 70 percent below the EPA standards for model year 1991 gasoline-fueled heavy-duty engines.[10]

Chlorofluorocarbons. Chlorofluorocarbons (CFCs) have been identified as major contributors to the breakdown of the Earth's protective ozone layer, and are thought to be effective greenhouse gases. While atmospheric concentrations of CFCs are relatively small, they are by far the fastest growing of the greenhouse gases on a percentage basis.[11] Most natural gas systems do not use chlorofluorocarbons.

Carbon Monoxide. High concentration levels of carbon monoxide (CO) can cause serious illness or, in extreme instances, death. The incomplete combustion of carbon-based fuels produces carbon monoxide, a colorless and odorless gas. The concentration of CO in the atmosphere is dependent on time of day, location, weather and human activities. CO levels tend to be highest in areas of heavy vehicular traffic. When concentrations exceed 100 parts per million, nausea, headache and dizziness are common.

Nearly three-fourths of all CO emissions stem from the operation of vehicles. The use of natural gas in place of other vehicular fuels can significantly reduce carbon monoxide emissions. For example, the EPA conservatively estimates that natural gas vehicles can yield a 50 percent reduction in carbon monoxide emissions.[12] A recent study at Princeton University found that with lean mixture control systems, carbon monoxide emissions "can be effectively eliminated from the exhaust of natural gas-fueled engines."[13]

Reactive Hydrocarbons. Photochemically reactive hydrocarbons react in sunlight with other air pollutants to form smog, which can irritate the nose and eyes, reduce visibility and lung function, aggravate respiratory diseases, and corrode and soil fabric. The primary component (over 90 percent) of natural gas is methane, a non-reactive hydrocarbon which does not contribute to smog formation.

Gasoline- and diesel-powered transportation vehicles and various petroleum operations are responsible for the majority of man-made hydrocarbon emissions. The EPA has previously estimated that use of natural gas in vehicles in lieu of gasoline can result in a 40 percent reduction in tailpipe emissions of reactive hydrocarbons (i.e., hydrocarbons that play a significant role in ozone formation) and a 100 percent reduction in emissions of reactive hydrocarbons from filling stations and evaporation.[14]

More recent data indicate that use of natural gas as a vehicular fuel can reduce reactive hydrocarbon emissions in total (tailpipe plus evaporative and filling) by over 80 percent.[15] Total hydrocarbon emissions (non-reactive, primarily methane, plus reactive) in today's vehicle are approximately equal car-for-car when using either gasoline or natural gas. In vehicles designed and built to use natural gas, however, even total hydrocarbon emissions can be reduced 33-50 percent.[16]

Methane. Methane is often referred to as one of the fastest growing greenhouse gases. The growth rate of methane is considered to be 1 percent per year, roughly two-to-three times the percentage growth rate of carbon dioxide. The total concentration of methane in the atmosphere, however, is only about 1/200 that of carbon dioxide, and it has a relatively short average atmospheric life of only 11 years. Consequently, the absolute growth rate of the number of molecules of methane in the atmosphere is only 1/74 that of CO_2.

Many analysts consider a molecule of methane to be 20-30 times more effective as a greenhouse gas than a molecule of carbon dioxide, although this point is in dispute. Assuming that methane actually is 30 times more effective as a greenhouse gas than carbon dioxide, and considering a number of other factors, analysts conclude that methane accounts for approximately 15-20 percent of the future greenhouse problem.

Producing and using natural gas reduces methane emissions from landfills and coal seams, as well as seepage from natural gas reservoirs.[17] These reductions appear to be large, but are not yet well quantified, so that it is not clear whether natural gas industry operations reduce or increase net methane emissions. In the US, known venting of methane from natural gas industry operations equals only half of known venting from coal seams and as little as 5 percent of methane venting from waste landfills.[18,19] At the worst, however, when ignoring the reductions side of the equation entirely, worldwide natural gas industry opertions appear to account for 5 to 8 percent of total worldwide methane emissions and about 1 percent of the postulated greenhouse effect.[7,20,21,22,23,24]

Non-Combustible Solid Waste and Sludge. Natural gas combustion results in no ash or sludge production. According to the US Environmental Protection Agency, a 1,000 MW powerplant operating on high-sulfur coal with flue gas desulfurization would produce 700,000 tons of sludge per year and 250,000 tons of ash. A unit of the same size operating on 3 percent sulfur residual oil with a scrubber would produce no ash, but would produce 450,000 tons of sludge annually. (Oil desulfurization at the refinery is an option to scrubbing at the plant site.) There would be no production of either sludge or ash if the plant were to operate on natural gas.[25]

Dissolved Solids. Water-dissolved solids discharges from natural gas industry operations are neglible. For example, coal mining, processing and combustion account for approximately one-half of all dissolved-solids discharges in the US, versus less than 5 percent attributable to natural gas operations.[26]

Water Consumption. Water consumption associated with natural gas use is generally far lower than for other energy alternatives, which require greater processing or conversion. Even when natural gas is used for electric power generation using combined-cycle technology, total treated water consumption by the gas-fired plant is only half that associated with coal-fired electricity generation.[27]

Capital Efficiency and
Resource Advantages of Natural Gas

As a rule, supply and utilization systems based on gaseous fuels require substantially less initial capital investment than alternative energy systems. For example, supplying added quantities of gaseous fuel from conventional lower-48 US gas resources for residential space heating would require only 29 percent to 46 percent of the initial capital required for the same amount of useful heat provided by nuclear or coal-fired electric power.[28]

A major factor contributing to the capital efficiency of gas systems is their high thermal efficiency. For example, a unit of natural gas can be produced and delivered to the end-user at an efficiency of nearly 89 percent. By comparison, when coal is used to generate electricity, the efficiency of generation and delivery is only 26-28 percent -- that is, more than 70 percent of the resource energy is wasted. Even when a high-efficiency electric heat pump is used, the total efficiency of the electric cycle is 48 percent as compared to 85 percent for the gas system using a high-efficiency furnace.

Natural Gas
Applications and Technologies

This section of the chapter provides an introductory overview of some of the major gas utilization applications and technologies which are emerging, with particular emphasis upon environmental benefits. More detailed discussion of each follows later in the chapter.

Natural Gas-Fueled
Combined-Cycle Electric Power Generation

Whereas two-thirds of the energy input to conventional fossil fuel boiler systems is normally lost to the environment (30 to 34 percent efficiency), the waste heat from the gas turbine in a combined-cycle unit is captured and utilized. Thus, combined-cycle system efficiency is in the 45 percent range or above.

A major advantage of combined-cycle systems is their relatively low capital cost -- approximately one-third the cost per kilowatt of coal-based capacity, and about one-fifth the cost per kilowatt of nuclear capacity. The cost of combined-cycle systems, most of which are largely pre-packaged, has risen far less rapidly than coal and nuclear units. A recent study found that in the United States a new 240 megawatt (MW) combined-cycle unit produces electricity at a cost which is only 54 to 92 percent of the cost of electricity produced by a new, comparably sized coal unit.

In addition, combined-cycle systems can be brought on-line quickly and in stages. For example, two-thirds of the electric capacity of a 100 MW combined-cycle system (the gas turbine component) can be on-line in 12 to 18 months, with the balance operating in an additional 6 months. In contrast, new coal or nuclear units require approximately 8-15 years for design and construction.

The environmental advantages of combined-cycle plants are also significant. For example: (1) a gas-fired combined-cycle unit emits less than 1 percent of the SO_2 and particulates associated with an equivalent coal-fired unit; (2) emissions of NO_x from the combined-cycle plant are 10-57 percent of the coal unit emissions; (3) the coal plant would produce 60 thousand tons of ash per year and 168 thousand tons of sludge while combined-cycle units produce neither of these non-combustible solid wastes; and, (4) treated water consumption by the combined-cycle plant would be 15 million gallons per year, one-half the consumption of the coal plant.[29]

For example, if 100 electric power plants totaling 24,000 MW were to be built instead of the same amount of new coal-fired electric capacity with full new source performance standard (NSPS) controls, annual SO_2 emissions would be 410,000 tons lower and annual NO_x emissions would be reduced by 105,000 tons (43 percent). Total annual natural gas use by these 100 plants would be about 1 TCF. Annual CO_2 emissions would be 143,000 tons if the plants were coal-fired, compared to 51,000 tons for combined-cycle gas (See Table 3.2).

Natural Gas-Fired Cogeneration

Whereas combined-cycle units (see above) utilize waste heat to generate electricity, cogeneration systems use waste heat for a variety of purposes -- e.g., space or water heating and industrial process needs.

Gas-fired cogeneration systems are an attractive option from both an environmental and an energy efficiency standpoint. According to the US Office of Technology Assessment, a gas turbine cogeneration system would require approximately 25 percent less input energy than the combination of a new coal-fired electric powerplant and an oil-fired boiler producing steam (assuming optimal system sizing). In addition, the gas-fired cogenerator would emit less than 1 percent of the SO_x, 27

56

Table 3.2. Emissions Resulting from 100 New 240 MW Gas Combined Cycle Powerplants vs. Coal (Annual Tons)

	(A) Combined Cycle	(B) NSPS* Coal	(C) Difference (B-A)	Un-- controlled Coal	Difference (C-A)
Sulfur Dioxide	0.3	410	410	3,900	3,900
Particulate Matter	0.9	21	20	1,700	1,700
Nitrogen Oxides	135	240	105	680	545
Carbon Dioxide	51,000	143,000	92,000	143,000	92,000
Gas Input/Year	1.08 Quad				

*NSPS: A unit meeting U.S. New Source Performance Standards.

percent of the particulates, and 50 percent of NO_x produced by the comparably sized conventional coal- and oil-based system (with pollution control equipment).

Gas-fueled cogeneration is also economically and environmentally preferable to coal-fueled cogeneration in large-scale industrial and commercial applications. For example, when compared to a coal-fueled system with a scrubber and an electrostatic precipitator, a gas-fueled steam turbine emits only 6 percent of the sulfur oxides, 10 percent of the particulates, and 28 percent of the nitrogen oxides.[30,31]

Natural Gas Repowering

By 1990, approximately 25 percent of all fossil-fueled generating capacity will be at least 30 years old. As generating equipment ages, it presents plant operators with a variety of problems, including reduced efficiency and reliability. A 30-year-old unit requires 6 to 8 percent more fuel -- at a cost of millions of dollars annually -- to generate the same amount of power as it did when it was first put into service. Unit reliability also suffers over time. A 10-year-old coal-fired plant has a forced outage rate of about 5 percent; however, this rate increases to nearly 15 percent within 30 years.

Combined-cycle repowering refers to the integration of new and used equipment at an existing site, with the final equipment configuration resembling a new gas-fired combined-cycle unit.

The type of repowering will vary from site to site. For example:

- Peaking turbine repowering refers to the addition of a steam turbine and heat-recovery unit to an existing gas turbine, with the efficiency improvement allowing the unit to convert from peaking to baseload operation.

- Heat-recovery repowering is the replacement of an old coal boiler with a new gas turbine and heat-recovery unit, leaving the existing steam turbine in place.

- Boiler repowering feeds the exhaust from a new gas turbine into an existing coal boiler, replacing the existing coal boiler's forced draft fans and combustion air heaters, and significantly reducing the system's electrical and fuel consumption.

The cost of producing electricity in a repowered unit is only about 60 percent of the cost of producing electricity in a new, coal-fired power plant. Even if the delivered price of natural gas is substantially higher than the price of coal, this price differential is likely to be more than offset by the lower capital and operating costs and maintenance expenses associated with the repowered system. The capital charge for a repowered unit is very low -- between 0.6 cents and 1.0 cents per kilowatt hour (kWh), versus 3.3 cents/kWh for a new coal unit -- since a repowered unit uses some existing equipment and does not require pollution-control devices. A related benefit is the ability to keep the power-generating facility on the existing site, eliminating the need for acquiring additional land, rights-of-way and various permits required for new construction.[32]

Another important economic factor is that the fuel cost advantage enjoyed by coal units is further diminished because of the 10 to 30 percent higher efficiency of repowered gas units.

Similarly, the operating and maintenance expenses of repowered units are only about one-third that of new coal plants, which require coal storage and handling, as well as the maintenance of pollution-control equipment and sludge and ash disposal.

Fuel Switching and Select Use in Boilers

Switching to gas can be a highly attractive and cost-effective control technique in existing boilers.

- Using gas in conjunction with, or in place of, coal or residual oil entails minimal up-front capital costs and relatively low life-cycle costs. Many large coal and oil boilers currently have the capability to also burn gas. Those that do not can add this capability at a very low cost -- $5 to $20 per kW -- with little or no loss in boiler rating or efficiency.[33,34]

- The cost of removing SO_2 via "select use of gas" ranges from $50 to over $1000 per ton, often in the $200 to $400 range. Alternative control strategies range from $250 to $2000 per ton removal. NO_x control may be thought of as a free extra gas benefit.[35]

- As noted above, switching to gas reduces a number of pollutants simultaneously. In addition, it offers quick response and supply security.

- Using gas can also be used in conjunction with other control strategies, such as high-sulfur coal washing, thereby allowing the continued use of local coals. For example, SO_2 reductions of 50 to 66 percent could be achieved while maintaining the local coal as 70 percent of the total fuel mix.

The substitution of 1 Tcf/year of natural gas for high sulfur coal in boilers would have a significiant impact on a number of pollutants.

- SO_2 and NO_x emissions would be reduced by 2.8 million tons and 0.5 million tons per year, respectively. By combining select use with coal washing, the SO_2 reduction could be roughly doubled.

- In addition, non-reactive hydrocarbons would be reduced by 2.3 thousand tons per year, and CO would be reduced by 5 thousand tons per year.

An A.G.A. study also investigated the relative advantages of natural gas boilers and fluidized bed coal boilers.[36]

For a typical industrial application, the use of natural gas in an boiler costs only 73 percent of the cost of burning coal in an atmospheric fluidized bed combustion (FBC) boiler, while emitting only a fraction of the air and solid waste pollution used by the FBC system. While the fuel costs of the gas system are higher, the gas boiler's lower capital and non-fuel operation and maintenance expenses more than offset the fuel cost advantage of the FBC system.

Eight alternate cases evaluated economic sensitivity of changing various parameters, including: unit size, fuel prices, capacity utilization, discount rates, and capital costs. In all cases, the gas option was found to cost less, ranging from 64 percent to 86 percent of the FBC unit. However, the option of burning coal in a conventional boiler equipped with a scrubber was found to be more costly than the FBC boiler.

Although the combustion process of an FBC boiler reduces the sulfur dioxide and nitrogen oxides emissions associated with uncontrolled coal burning, and usually employs a fabric filter to control particulate emissions, air pollution emissions are much greater for the FBC boiler than the gas boiler. For example, the natural gas system emits less than one percent of the sulfur dioxide, only 24 percent of the nitrogen oxides and 11 percent of the particulate matter compared to a similarly sized coal FBC.

Cofiring

Gas cofiring involves simultaneously burning natural gas -- typically comprising less than 30 percent of boiler fuel input -- with coal or oil in the same boiler to reduce emissions and improve operational performance.

Cofiring gas with coal in the same boiler offers a variety of environmental and operational benefits. Not only does the use of gas reduce emissions of SO_2, NO_x, particulate matter, sludge, ash and many other pollutants, it also can enhance boiler performance, resulting in lower maintenance costs and fewer plant outages. Cofiring also provides cleaner and quicker start-ups.

Recent field experience has documented a greater than one-to-one reduction in SO_2 and NO_x emissions with cofiring. Theoretically, for SO_2, the reduction should be directly proportional to the amount of coal displaced. Using gas to provide 10 percent of the fuel input, for example, should result in 10 percent less SO_2. However, in practice, GRI research showed that measured levels reflected significantly greater reduction, apparently due to the tendency of the sulfur to bind to the ash during the process. Related tests have shown that NO_x emissions are reduced at a rate roughly three times greater than the proportion of gas used.[37]

Seasonal Gas Substitution

Natural gas also can be used to reduce SO_2 emissions by substituting or alternating its use with coal or high-sulfur oil. Controlling SO_2 emissions during specific periods of the year may be more cost-effective than year-round controls, as indicated in studies conducted by the Massachusetts Institute of Technology (MIT). According to MIT, 70 percent of annual acid deposition nationwide falls from May through October, with 54 percent falling during the four months of June through September. This coincides with the period of greatest available natural gas pipeline capacity. Substituting gas for dirtier fuels also can provide very cost-effective compliance under EPA's emissions trading program.[34,38]

Gas Reburning with
Sorbent Injection

In gas reburning, natural gas replaces a portion of the coal burned in a boiler. The gas, injected into the upper region of the furnace, creates a fuel-rich zone in which NO_x is chemically changed into a harmless molecular nitrogen. Research conducted by GRI and Energy and

Environmental Research Corp. show that reburning could reduce NO_x emissions by 50 to 60 percent.[39]

Sorbent injection involves mixing a sulfur-absorbing chemical with the coal to remove the sulfur oxide gases that are formed when coal burns. According to Energy and Environmental Research, use of sorbent injection is expected to cut SO_2 emissions by at least one-half.

Vehicular Applications

Conversion of gasoline or diesel fueled vehicles to natural gas can reduce carbon monoxide emissions by up to 99 percent, reduce nitrogen oxides emissions by up to 65 percent, and/or reduce reactive hydrocarbon emissions by up to 92 percent. This is particularly true for trucks, some buses, and older cars, which account for 60-90 percent of vehicular emissions. Actual emissions reductions achieved depend upon a large number of factors including tuning, age and design of the vehicle, and the condition and technology of the gasoline emissions controls and natural gas conversion kit.[40]

Residential Gas Systems

Natural gas has long been recognized to be the most economical fuel for residential space heating, water heating, clothes drying and cooking. Recent commercialization of a new generation of highly efficient gas appliances has enhanced this advantage.[41,42]

It is less often recognized that, as with other gas uses, residential gas systems are also best from the perspective that they utilize less primary energy and result in less environmental pollution, than do most alternatives.

Natural Gas Cooling Systems
("Air Conditioning")

The recent commercialization of new, high-efficiency gas cooling systems has greatly improved the attractiveness of natural gas for this application. American Gas Association analyses have demonstrated the cost advantages of gas cooling in many commercial applications. Perhaps even more important from an energy planning and environmental perspective, the availability of gas cooling may avoid the need for adding highly expensive electric peaking capacity to meet summer loads, while also balancing seasonal gas demand.[43,44,45]

PART II: DETAILED ASSESSMENTS
OF GAS APPLICATIONS

This part of the chapter summarizes a number of more detailed assessments of environmental impacts and benefits of natural gas use related to specific gas technologies. Most costs are stated in US dollars for facilities located in the US, but the conculsions are believed to be illustrative and directionally correct for many situations.

An Evaluation of Alternative Control Strategies to Remove Sulfur Dioxide, Nitrogen Oxides and Carbon Dioxide at Existing Large Coal-Fired Facilities

The primary precursors of acid rain are believed to be sulfur dioxide (SO_2) and, to a lesser extent, nitrogen oxides (NO_x). The bulk of these two pollutants are released into the atmosphere by the combustion of some fossil fuels in stationary sources (e.g., industrial boilers and electric utility powerplants) and mobile sources (e.g., automobiles). Total annual US emissions are in the 19 million ton range for SO_2, and 17 to 18 million tons for NO_x. Nearly three-fourths of all SO_2 emissions and one-third of all NO_x emissions are attributable to the combustion of coal at large powerplants.

A distinct but related environmental issue is that of global warming or the "greenhouse effect." The greenhouse effect refers to the trapping of solar heat in the earth's atmosphere resulting from a build-up of "greenhouse gases," particularly carbon dioxide (CO_2). CO_2, similar to SO_2 and NO_x, is generated primarily by fossil fuel combustion.

This section of the chapter summarizes the results of a study of the costs of removing SO_2 and NO_x emissions from large, existing coal-fired facilities via alternative control strategies. (CO_2 removal was not included in the cost analysis because CO_2 emissions dwarf NO_x and SO_2 emissions and would dominate the analysis.) In addition, the potential SO_2, NO_x and CO_2 reduction levels attainable through the various strategies are quantified.

Ancillary costs and benefits attributable to the various strategies --such as the production of scrubber sludge, ash, or other solid wastes, or air or water impacts -- are not considered in this analysis. It should be noted, however, that ancillary impact assessments tend to heavily favor gas-based options.

Of the 15 control strategies analyzed, repowering with natural gas in either a boiler conversion or combined-cycle repowering offers the greatest potential reduction in pollutant emissions (see Table 3.3). Further, only the five natural gas options simultaneously reduce each of the three pollutants considered in this analysis -- SO_2, NO_x and CO_2. From an economic perspective, the gas strategies are extremely competitive, with combined-cycle repowering being the least-cost baseload option among those strategies that offer appreciable reduction levels.

Each of the control strategies was compared on the basis of the combined cost-per-ton of SO_2 and/or NO_x removed from a 500 megawatt existing coal-fired powerplant in constant levelized 1988 dollars. Three cost scenarios -- based on low, high and mid-range cost assumptions -- were run for both baseload (65 percent capacity) and intermediate load (30 percent capacity) facilities.

Natural Gas Boiler and Combined-Cycle Repowering. Repowering with natural gas in a combined-cycle mode can reduce emissions of SO_2, NO_x and CO_2 per unit of electricity generated by 100, 90 and 64 percent, respectively. Converting to gas in the existing boiler also provides significant reductions in each of the three pollutants -- 100, 45 and 43 percent, respectively -- although NO_x and CO_2 reductions are not as great because of the greater efficiency of combined-cycle systems.

Primarily attributable to its very high removal efficiency, combined-cycle repowering is the least-cost option in baseload situations among those strategies that offer appreciable reductions -- \$180 per ton of SO_2 and NO_x removed (mid-range cost scenario). Boiler repowering is one of the lowest cost options in intermediate load situations -- \$315 per ton -- as the very low ratio of fixed to total costs does not penalize this option at reduced capacity as much as most other options.

Gas Reburn and Reburn with Sorbent Injection. Reburning with a mixture of 20 percent gas and 80 percent coal is expected to reduce NO_x emissions by 60 percent, along with SO_2 reductions of 20 percent and CO_2 reductions of 9 percent. The mating of sorbent injection with reburn can increase the SO_2 reduction to 50 percent.

Gas reburn is a moderate cost strategy (\$325 per ton baseload, \$360 per ton intermediate load) while reburn with sorbent injection is a moderate to high cost strategy, as the enhancement of SO_2-capture significantly raises O&M expenses.

Gas/Coal Cofiring. NO_x reductions of 25 percent, along with SO_2 and CO_2 reductions of 12 and 4 percent, respectively, can be achieved by cofiring a mixture of 10 percent gas and 90 percent coal. There is no technical constraint that precludes using more than 10 percent gas, but testing to date has shown greater than 1:1 reductions at this gas level, which is particularly attractive.

Table 3.3. Emission Reduction Potential in Large Coal-Fired Boilers (Percent Reduction Achievable)

Control Technology	Sulfur Dioxide	Nitrogen Oxides	Carbon Dioxide
I. Precombustion			
Lower Sulfur Coal	70	-	-
Conventional Coal Cleaning	30	-	-
Advanced Coal Cleaning	50	-	-
II. Combustion			
Low Excess Air	$()^5$	15	$()^5$
Overfire Air	$()^5$	30	$()^5$
Low NO_x Burners	-	50	-
LIMB	55	55	-
Gas/Coal Cofiring[1]	12	25	4
III. Post-Combustion			
Selective Catalytic Reduction	-	80	-
Flue Gas Desulfurization	85	$(8)^6$	$(8)^6$
Sorbent Injection	60	-	-
Gas Reburn[2]	20	60	9
Gas Reburn + Sorbent Injection[3]	50	60	9
IV. Repower			
Natural Gas Conversion[4]	100	45	43
Combined Cycle Conversion	100	90	64

[1] 10 percent natural gas.
[2] 20 percent natural gas.
[3] 20 percent natural gas.
[4] Based on 100% conversion to natural gas.
[5] Minor decrease due to unit efficiency increase.
[6] Increase in emissions due to lower unit efficiency.

Excluding the three low-cost low NO_x removal strategies (see below), gas cofiring is the second most cost effective strategy in baseload conditions ($225 per ton) and the most effective in intermediate load ($245 per ton).

Flue Gas Desulfurization (Scrubbing). Scrubber retrofits, when technically feasible, can reduce SO_2 emissions by 85 percent. Both NO_x and CO_2 emissions will increase by about 8 percent, however, because of the electricity needs of the scrubber that reduce system efficiency.

Scrubbers are a moderate to high cost strategy under mid-range cost and baseload assumptions ($470 per ton), and are the second most costly strategy assuming intermediate load ($765 per ton, second only to selective catalytic reduction). Scrubbing is dissimilar to most other strategies in that it has both high capital and high O&M expenses.

Low Sulfur Coal and Coal Cleaning. Coal cleaning offers moderate level SO_2 reductions (30 percent with conventional methods, 50 percent with advanced), while converting to low sulfur coal can reduce SO_2 by 70 percent. (This analysis based on converting from 3.5 percent to 1.0 percent sulfur coal.)

Converting to lower sulfur coal is a moderate cost option at $370 per ton baseload. Advanced coal cleaning is a relatively low cost option -- $275 per ton baseload -- but conventional coal cleaning at $415 per ton is significantly more costly because of its lower SO_2 removal efficiency.

Selective Catalytic Reduction (SCR). SCR can reduce NO_x emissions by 80 percent (although unproven for many US coal and boiler types), but is roughly 10 times as costly as the average cost reduction strategy.

Low Excess Air, Overfire Air, Low NO_x Burners. These three combustion modification strategies offer low to moderate level NO_x reductions (15, 30 and 50 percent) at low cost ($20 to $265 per ton). They are not feasible in all boiler types, however, and would probably have to be augmented with other strategies to reach acceptable reduction levels.

It must be remembered that retrofitting coal-fired powerplants is a site specific proposition, and even the broad ranges presented in this analysis should not be construed as the possible extremes. For example, the scrubber retrofit in a high cost intermediate load situation presented herein has a removal cost of $1,000 per ton. If, however, we assumed the base unit was initially operating on 2 percent sulfur coal as opposed to 3.5 percent sulfur coal, the removal cost would jump to over $1,750 per ton as less SO_2 would be removed at the same cost. The attempt here is to present consistent comparisons under uniform conditions. It is clear, however, that virtually all the strategies examined in this analysis (with the exception of SCR) could be attractive given particular emission reduction goals and site specific parameters -- e.g., relative fuel costs, unit age and design, and space availability.

Background and Description of Control Options. Most proposed or implemented efforts to curb acid rain focus on the reduction of SO_2 and NO_x at existing coal-fired electric utility powerplants. Post-1978, or "new" electric utility powerplants, are governed by US federal new source performance standards that require a 70 to 90 percent reduction in SO_2 from uncontrolled levels and a 65 percent reduction in NO_x at coal-fired units. These sources are therefore usually excluded from consideration of further reduction.

65

There are a number of ways to reduce the pollutant emissions from large existing coal-fired boilers. For purposes of this analysis, the various reduction strategies have been grouped into four general categories. *Precombustion* options refer to the cleaning of the existing coal type prior to combustion, or the purchase of a new and cleaner coal type. *Combustion* options refer to modifications in the combustion chamber that result in lower levels of pollutant emissions from the combustion process. *Post-Combustion* options are those that somehow trap or remove the pollutant products from combustion before they exit the stack into the atmosphere. *Repowering*, in this analysis, refers to the conversion to a different fuel (e.g.,. coal to natural gas) at the same plant site in either new or existing combustion equipment. Each of the control options analyzed in this analysis is briefly described below.

Conversion to a *Lower Sulfur Coal* is a sometimes easy and economical means of reducing SO_2 emissions, although it does not affect NO_x or CO_2 emissions. Coal consumed by US electric utilities ranges from less than 1 percent sulfur to more than 5 percent. This analysis is based on a conversion from 3.5 percent sulfur to 1.0 percent sulfur coal, which would reduce SO_2 emissions by 70 percent (see Table 3.3). Such conversions may or may not be difficult depending on the specific facility and coals involved. Converting from US Eastern high sulfur to US Western low sulfur coal may require boiler modifications to avoid slagging and fouling, and precipitators and coal handling equipment may have to be expanded to handle the greater quantities of low sulfur coal required -- since low sulfur coal typically has a lower Btu content than high sulfur coal. For example, US Western coal, on average, has about one-fourth the sulfur content of Appalachian coal, but it also has about 25 percent fewer Btu per pound.[46]

Coal contains sulfur in two forms -- pyritic and organic. Physical coal cleaning (i.e., grinding, breaking and washing) can reduce the pyritic sulfur in coal, but only chemical cleaning -- which will likely not be available until beyond the year 2000 -- can reduce the organic bound sulfur in coal. *Conventional Coal Cleaning* can reduce SO_2 emissions by 10 to 30 percent. *Advanced Coal Cleaning*, which involves more sophisticated physical techniques such as electrostatic separation, multi-stage floatation and oil agglomeration, is currently being demonstrated, and SO_2 reductions of 35 to 65 percent are expected.[47]

Three combustion modification techniques -- *Low Excess Air*, *Overfire Air* and *Low NO_x Burners* -- are relatively easy means to achieve low to moderate level NO_x reductions. Low excess air and overfire air are basically tuning techniques -- the former reduces the oxygen used in combustion, while the latter redirects some of the combustion air above the

66

top row of burners. Low excess air typically reduces NO_x by about 15 percent as compared with about 30 percent for overfire air.[48] These techniques may slightly improve unit efficiency, thereby resulting in minor CO_2 and SO_2 reductions. Low NO_x burners are a new breed of burners that seeks to provide a lower temperature combustion zone, thereby reducing NO_x emissions by some 50 percent.[49] These burners, which have limited retrofit experience and cannot be used in all boiler types, do not affect SO_2 or CO_2 emissions.

LIMB is the acronym for limestone injection multi-stage burner, a technology that seeks to reduce both SO_2 and NO_x emissions by 50 to 60 percent. The hope has been to mate two separate technologies -- limestone injection for SO_2 control and low NO_x burners for NO_x control. A limestone sorbent is injected with the coal for sulfur-capture through the formation of calcium sulfate, which can be collected and removed; the low NO_x burner is similar to that described above. Tests to date have indicated problems with fly ash collection, boiler fouling and corrosion. Movements away from limestone to some other type of sorbent are likely. Testing has also indicated that LIMB will not be feasible in all units, possibly restricted to smaller and older units.[50]

Gas/Coal Cofiring refers to the injection of natural gas with pulverized coal into the primary combustion zone of a boiler. Since natural gas contains no sulfur or fuel bound nitrogen, and less carbon than coal, emissions of these pollutants are reduced as gas replaces some portion of coal in the fuel mix. Cofiring is not a new technology, but efforts to determine optimal natural gas injection levels for both environmental and operational benefits (e.g., reduce precipitator fouling and smoky start-ups, and enhance the combustion of lower quality or wet coal) are relatively recent. Testing at the 570 megawatt Cheswick powerplant (Pennsylvania) indicate that a gas:coal ratio of 10:90 can reduce NO_x and SO_2 by 25 and 12 percent, respectively.[51] Excess air was reduced in this testing. This mixture would also reduce CO_2 emissions by some 4 percent.[52]

Selective Catalytic Reduction (SCR) is a post-combustion NOx control technology that does not affect SO_2 or CO_2 emission levels. SCR involves the use of a catalytic reactor and ammonia to reduce NO_x to nitrogen. SCR can achieve high level NO_x reductions -- 80 percent -- but it is a costly technology with serious concerns over system reliability, catalyst life, waste disposal and the effects of ammonia by-products on plant components. SCR has been demonstrated in some boiler types in Japanese and European utilities, and pilot testing is underway in the US.[53]

Flue Gas Desulfurization or "scrubbing" is a process whereby the gases produced by coal combustion are sprayed with water and an alkaline reagent, usually limestone, forming calcium sulfite or sulfate, that can be removed. Scrubbing is a proven technology that can remove up to 95 percent of uncontrolled SO_2 emissions, although slightly lower removal levels -- 75 to 90 percent -- would be likely in retrofit applications. Nearly 150 electric utility powerplants in the US have scrubbers today, accounting for roughly 20 percent of total coal-fired electricity generating capacity.[54,55] Scrubbing can be very difficult (i.e., costly) or impossible in some retrofit situations, however. In addition, scrubbers produce about 1 acre/foot of sludge per MW per year, and system efficiency is decreased by 6 to 10 percent with scrubbers thereby forcing increased coal throughput with increased NO_x and CO_2 emissions.[56]

The injection of lime, limestone, sodium carbonate or some other sorbent for SO_2-capture is referred to as *Sorbent Injection*. Sorbent injection (basically the LIMB technology discussed above without the NO_x control) can reduce SO_2 emissions by about 60 percent.[57]

Gas Reburn refers to the injection of natural gas into the upper furnace region of a boiler to produce a fuel-rich zone thereby reducing NO_x. Testing indicates that a 20:80 mixture of gas:coal can reduce NO_x and SO_2 by 60 percent and 20 percent, respectively.[58] In addition, CO_2 emissions would be reduced by about 9 percent.[59] Gas Reburn/Sorbent Injection mates the two previously discussed technologies in an effort to increase the SO_2 removal from the 20 percent level to 50 percent.[60] (50 percent SO_2 reduction is the current demonstration project goal, but it is not necessarily the maximum reduction achievable.) An attractive feature of the reburn technologies is that they are easily retrofitable and compatible with all boiler types, including cyclone boilers that are relatively high-level NO_x emitters.

High-level reductions in all three of the pollutants under consideration can be achieved via "site repowering." Site repowering maintains the existing plant site, but natural gas is substituted for coal either in the existing combustion equipment with minor burner modifications (*Natural Gas Conversion*) or in new and more efficient equipment (*Combined-Cycle Conversion*). Many powerplants are already dual-fuel capable, and conversions to gas on a seasonal or permanent basis in such units are relatively easy. In units not so equipped, conversions are straight-forward, but gas transmission facilities may or may not be nearby. Converting a boiler from coal to gas can reduce SO_2, NO_x and CO_2 by 100, 45 and 43 percent respectively.[61] Even greater reductions can be achieved by replacing the existing boiler with gas-fired combined-cycle equipment, which would reduce fuel requirements by about one-third -- 100 percent SO_2 reductions, 90 percent NO_x and 64 percent CO_2.[62] Combined-cycle equipment is readily available "off the shelf."

Recommendations

There are a number of conclusions to be drawn from this analysis.

- Of the 15 control technologies analyzed, six can reduce both SO_2 and NO_x -- five of these six use natural gas (which would also reduce CO_2 emissions).

- NO_x is significantly more expensive to control than SO_2, except for control strategies capable of providing only low level NO_x reductions, such as low excess air or overfire air.

- Selective catalytic reduction is by far the most costly control technology analyzed -- about five times the cost of the next most costly technology, and 10 times the cost of the average technology in most scenarios.

- Converting units to natural gas, either in an existing boiler or a new combined-cycle unit, is by far the most effective means analyzed of reducing pollutant emissions. Further, these gas options are extremely competitive from a cost perspective relative to the other options.

- NO_x is significantly more expensive to control than SO_2, except for control strategies capable of providing only low level NO_x reductions, such as low excess air or overfire air.

- Selective catalytic reduction is by far the most costly control technology analyzed -- about five times the cost of the next most costly technology, and 10 times the cost of the average technology in most scenarios.

- Converting units to natural gas, either in an existing boiler or a new combined-cycle unit, is by far the most effective means analyzed of reducing pollutant emissions. Further, these gas options are extremely competitive from a cost perspective relative to the other options.

- Those technologies that have a high ratio of capital costs to total cost will become relatively more expensive as unit capacity is reduced from baseload operation. Technologies that have a higher ratio of fuel costs to total cost (e.g., natural gas boiler conversion) may increase very little in terms of dollars per ton of pollutant removed as unit capacity is decreased.

Natural Gas-Fueled Combined Cycle
Electric Power Generation --
An Attractive Option

New natural gas-fired combined cycle units offer an efficient and economical alternative to electricity produced by new nuclear or coal-fired units. In addition, combined cycle power generation is attractive from an environmental perspective, with minimal impacts in terms of air pollutant emissions, solid waste generation and water consumption.

A comparison of the environmental impacts of a new 240 megawatt (MW) coal-fired electric powerplant versus a comparably sized natural gas-fired combined cycle unit has documented these advantages.

The environmental impacts of the alternative generating systems operating at a 65 percent capacity factor were estimated for three primary air pollutants -- sulfur dioxide (SO_2), particulate matter (TSP) and nitrogen oxides (NO_x). It was assumed that currently applicable federal emission standards for these air pollutants would be met. In addition, water consumption and the production of sludge and ash were also calculated. The coal system was assumed to operate on a commonly available coal type -- 2 percent sulfur, 10 percent ash -- with a flue gas desulfurization (scrubber) system for SO_2 control, and an electrostatic precipitator for TSP control.

In the US, a new coal-fired electric utility powerplant is subject to the new source performance standards of 1978 (NSPS) as set out in 40 *CFR* Subpart Da.[63] The NSPS for SO_2 mandates a sliding scale for SO_2 control: facilities emitting less than 0.60 pounds of SO_2 per MMBtu must have reduced potential (uncontrolled) emissions by at least 70 percent, while facilities emitting from 0.60 to 1.20 pounds per MMBtu must have reduced potential emissions by 90 percent. The scrubber employed in this analysis was assumed to be 85 percent effective, reducing uncontrolled emissions from 3.8 pounds of SO_2 per MMBtu combusted to 0.6 pounds per MMBtu.

The particulate NSPS is 0.03 pounds per MMBtu (99 percent reduction), and the NO_x NSPS requires a 65 percent reduction in total potential emissions with a cap of 0.60 pounds per MMBtu. The controlled coal plant emissions: SO_2 -- 0.60 pounds per MMBtu; TSP -- 0.03 pounds per MMBtu; and NO_x -- 0.35 pounds per MMBtu, were multiplied by the total annual coal input (13.7 trillion Btu) to obtain annual emissions. Uncontrolled air pollutant emissions were obtained from the US Environmental Protection Agency's (EPA's) *Compilation of Air Pollutant Emission Factors*.[64]

The only air emissions from the combined cycle unit originate in the gas turbine portion of the system, since there is no supplemental firing of the steam turbine. SO_2 and NO_x emissions from stationary electric utility gas turbines in excess of 100 MMBtu per hour and built after 1982 are subject to NSPS, but there is no standard for particulate emissions. The NO_x standard is approximately 75 parts per million of NO_x, which equates to 0.25 to 0.30 pounds of NO_x per MMBtu. SO_2 is limited to 0.015 percent of the exhaust gas total volume, but this standard is far in excess of actual SO_2 emissions from a gas turbine. Thus, uncontrolled SO_2 emissions from a gas boiler (0.0006 pounds per MMBtu according to US EPA, *Compilation of Air Pollutant Emission Factors*)[65] were used to estimate the negligible SO_2 emissions from a turbine.

Uncontrolled TSP emissions from a gas turbine operating on distillate oil are 0.004 pounds per MMBtu according to: Radian Corp, *Emissions from Stationary Gas Turbines*,[66] and emissions would be lower when operating on gas. It was assumed that the reduction when operating on gas would be 61 percent based on uncontrolled TSP emission factors for gas turbines contained in EPA's *Compilation of Air Pollutant Emission Factors* (0.004 pounds per MMBtu times 39 percent = 0.0016 pounds per MMBtu when operating on gas).

Coal-fired electric powerplants produce a number of solid wastes including ash and sludge. Fly ash is collected by electrostatic precipitators or baghouses, while bottom ash is formed by fuel combustion. Sludge is a semi-solid waste which results primarily from scrubbing for SO_2. These two non-combustible solid wastes must be disposed of in landfills or ponds, and according to the US EPA, *Energy/Environment Fact Book*,[67] a 1000 MW powerplant produces 700,000 tons of sludge per year and 250,000 tons of ash. These volumes were scaled down by 76 percent to estimate the wastes produced by a 240 megawatt unit.

One other environmental impact was considered in this analysis -- the consumption of treated water. Treated water consumption for the coal and combined cycle units was obtained from General Electric's *60 Hz STAG Combined Cycle Power Plants*.[68] Data cited in the report were for a 450 MW unit, and it were scaled down to a 240 MW unit assuming a linear relationship between plant size and water usage.

Environmental Results. Table 3.4 presents the environmental results of this analysis. As indicated, the gas-powered combined cycle system has negligible sulfur dioxide and particulate emissions. In contrast, the coal system emits over 4.1 thousand tons of SO_2 per year (with an 85 percent effective scrubber system) and 206 tons of particulate matter per year (with a 99 percent effective electrostatic precipitator). Emissions of NO_x by the

71

Table 3.4. Environmental Comparison of Natural Gas-Fired Combined-Cycle Generating Station and Coal-Fired Station (Annual Impacts, 240 Megawatts)

	Coal-Fired Steam Plant	Gas-Fired Combined Cycle
Air Pollutant Emissions (Tons/Year)		
Sulfur Dioxide	4,123[1]	3[2]
Particulate Matter	206[1]	9[3]
Nitrogen Oxides	2,385[1]	1,352[4]
Other Pollutant Emissions (Thousands of Tons/Year)[5]		
Sludge	168	none
Ash	60	none
Treated Water Consumption (Millions of Gallons/Year)[6]	30	15

[1]Based on new source performance standards for coal-fired electric powerplants (2% sulfur coal assumed): SO_2 -- 0.6 pounds per MMBtu, TSP -- 0.03 pounds per MMBtu, and NO_x -- 65% reduction. See: 40 *CFR* Subpart Da, Part 60.40a through 60.43a.

[2]Data on sulfur dioxide emissions from gas turbines not available, assumed to be equal to SO_2 emissions from gas boiler -- .0006 pounds per MMBtu. See: U.S. EPA, *Compilation of Air Pollutant Emission Factors*, (Research Triangle Park, N.C.: August 1982) P. 1.4-3.

[3] Gas turbines operating on distillate oil have particulate emissions of 0.004 pounds per MMBtu according to: Radian Corp. for the U.S. EPA, *Emissions from Stationary Gas Turbines, New Source Performance Standards*, (Research Triangle Park, N.C.: November 1984) P. 4-78. Emissions when operating on gas would be approximately 39% of this amount based on: U.S. EPA, *Compilation of Air Pollutant Emission Factors*, (Research Triangle Park, N.C.: August 1977) P. 3.3.1-2

[4]Based on new source performance standards for electric utility stationary gas turbines: NO_x -- 75 parts per million. See 40 *CFR* Subpart GG, Part 60.330 through 60.333.

[5]U.S., Environmental Protection Agency, Research and Development-Energy Minerals and Industry, *Energy/Environment Fact Book*, (Washington, D.C.: March 1978) P. 24.

[6]General Electric, GEA-11387, *60 Hz STAG Combined Cycle Power Plants*, (Schenectady, N.Y.) P. 568316A.

combined cycle system are 57 percent of the coal unit's annual NO_x emissions -- 1.352 tons per year versus 2,385 tons per year.

In terms of non-combustible solid wastes, the coal-based system would generate 168,000 tons of sludge per year and 650,000 tons of ash. Natural gas combustion results in no sludge or ash production. In addition, the combined cycle option would require only half the water of the coal system -- 15 million gallons per year versus 30 million gallons.

An Economic and Environmental Comparison of Natural Gas and Coal Use for Industrial Cogeneration

Cogeneration is the sequential production of power (electrical or mechanical) and thermal energy (steam or heat) within one system from the same fuel source. Since cogeneration processes employ the waste heat that is generated as a byproduct of the original power production procedure, they are usually more energy-efficient than conventional power production processes.

The higher energy efficiency of cogeneration plants, relative to conventional plants, allows them to simultaneously contribute to two major public-policy goals: energy security and a cleaner environment. Energy security emanates from the fact that the two major cogeneration fuels, natural gas and coal, are derived predominantly from domestic resources. A cleaner environment follows from increased energy efficiency. Additionally, gas-based systems benefit from the inherent cleanliness of natural gas.

From an environmental perspective, natural gas cogeneration, using either gas or steam turbines, is far superior to coal-based steam turbine cogeneration systems fitted with pollution control equipment. Table 3.5 compares the emissions from a gas turbine system (that includes a supplemental boiler) to two steam turbines systems, one fueled with natural gas and the other fueled with coal. All three systems have the same electric and thermal output (10,888 kWh and 100 tons/hr). Examining the two steam turbine systems reveals that gas-based steam turbines produce 94 percent less SO_2, 90 percent less TSP, 90 percent less NO_x and 43 percent less CO_2 than the coal-based steam turbine. The gas turbine, with a supplemental boiler, reduces SO_x 100 percent, TSP 85 percent, NO_x 80 percent and CO_2 42 percent relative to a coal-based steam turbine system.

Table 3.5 also illustrates the pollutant reductions from a gas turbine cogeneration system relative to a conventional industrial plant (i.e., one that purchases electricity from a central station coal-based power plant and produces its own steam from an oil-based boiler). Relative to the

73

Table 3.5. Annual Emissions from Conventional and Cogeneration System Operation[1,2] (Tons/Year)

Pollutant	Conventional Industrial[4] Plant	Steam Turbine Systems Coal-Based[5]	Gas-Based	Gas Turbine with Supplemental Gas Boiler Gas Turbine	Back-up Boiler	Total	% Reduction Achieved by Gas System vs. Coal System Gas-Based Steam Turbine	Gas Turbine System
SO$_x$	1,063	1,188	70	3	Neg[6]	3	94%	100%
TSP	91	100	10	10	5	15	90%	85%
NO$_x$	414	694	67	50	92	142	90%	80%
CO$_2$	206,479	187,658	106,835	45,132	65,780	110,912	43%	41%

[1] All systems produce 10,888 kWh and 200,000 lbs/hr of steam and operate at 70 percent capacity utilization.

[2] All emission factors, with the exception of NO$_x$ factors for gas-based steam turbines and gas turbines, are taken from the Office of Technology Assessment, Congress of the United States, *Industrial and Commercial Cogeneration* (Washington, D.C.: February 1983). The NO$_x$ standards of new cogeneration systems have been tightened significantly since 1983. This analysis assumes a mid-level control technology for gas-based steam turbines and gas turbines based on steam/water injection.

[3] Since emission calculations are based on equivalent electric and thermal output, a supplemental boiler is included in the gas turbine system.

[4] Conventional industrial plant that purchases power from NSPS coal plant and generates steam with oil-based boiler.

[5] Assumes the use of flue gas desulfurization and an electrostatic precipitator in order to meet New Source Performance Standards.

[6] Negligible.

conventional system, the gas turbine reduces SO$_2$ 100 percent, TSP 84 percent, NO$_x$ 66 percent and CO$_2$ 46 percent, while the gas-based steam turbine reduces SO$_2$ 99 percent, TSP 89 percent, NO$_x$ 84 percent and CO$_2$ 48 percent.

Although emission numbers are only presented for one system size, the emissions will increase in direct proportion with system size. Thus, if the system sizes are doubled so will the emissions. As such, Table 3.5 presents the relative reductions achieved when employing natural gas cogeneration over comparable coal-based cogeneration systems. Clearly, these lower emissions levels provide a significant supplemental bonus to the economic benefits provided by natural gas cogeneration systems.

Industrial Use of Natural Gas and Woodfuels: Competition and Co-Firing

Woodfuel is the generic term for a variety of byproducts of the paper, lumber, and furniture industries that can be burned in boilers and/or process heaters. Since woodfuel is an industrial byproduct, its procurement cost to those industries is often minimal. In addition, companies with excess woodfuel often sell the excess to other industries at a cost usually lower than that of most fossil fuels.

However, woodfuel consumption in industrial boilers results in significant owning and operating costs, even when the woodfuel is free. The non-fuel owning and operating costs of a natural gas-fired boiler are less than those of a woodfuel boiler, and when the impact of falling natural gas prices is considered, the fuel use decision between the two is not clear cut.

This section compares the economics and environmental impacts of burning woodfuel and natural gas in new boilers. The cost of burning woodfuel or gas in an existing dual-fueled boiler is also examined. In addition, the benefits of cofiring natural gas with woodfuels are illustrated.

Economic Comparison

New, Separate Boilers. A new natural gas boiler can be economically preferable to a new woodfuel boiler, even if the woodfuel is free. This seemingly implausible result stems from the high cost of owning and operating a woodfuel unit.

- The capital cost of a typical new woodfuel boiler ranges from 2.5 to 3.0 times the cost of a gas boiler with equal output.

- Operating and maintaining that woodfuel boiler, exclusive of fuel charges, costs from 1.7 to 2.4 times more than the gas boiler.

- Additionally, boiler use of gas is 23 to 33 percent more efficient than the use of woodfuel, requiring the purchase of larger (more expensive) woodfuel units.

When the above factors are considered, new natural gas boilers have a non-fuel owning and operating cost advantage ranging from $6.16/MMBtu for a small boiler to $3.29/MMBtu for a large boiler. Thus, as long as the price of natural gas is below this level -- $3.29/MMBtu to $6.16/MMBtu -- gas is economically preferable to boilers operating on "no cost" woodfuel. If there is a procurement cost for the woodfuel, including the cost of preparing the woodfuel and/or obtaining the woodfuel from an outside supplier, then the breakeven price for natural gas is increased.

Existing Woodfuel/Gas Boiler. An existing boiler capable of firing either woodfuel or natural gas will also experience lower non-fuel O&M costs when operating on natural gas. However, this cost advantage for natural gas operation is lower, ranging from $0.27/MMBtu for a large boiler to $0.56/MMBtu for a small boiler. The non-fuel cost advantage for gas in a single dual fuel capable boiler is less than the gas advantage in new, separate boilers because the capital recovery expenses and the fixed O&M costs are constant, regardless of the fuel burned.

While this lower cost advantage precludes natural gas from competing economically with free woodfuel, natural gas can compete with purchased woodfuels in existing woodfuel/gas boilers in some instances. For example, natural gas at $3.00/MMBtu would be competitive with woodfuels purchased for $2.00/MMBtu as a result of both the boiler efficiency difference between the two fuels and the non-fuel O&M cost differential.

Environmental and Operating Aspects

Despite the paper and lumber industries' attempts to displace fossil fuels with woodfuels wherever possible, these industries use a substantial amount of natural gas in boilers either alone or in a woodfuel mixture, due to either an insufficient woodfuel supply, the fuel use economics of new boilers, or combustion problems associated with operating solely on woodfuels. Almost all wood burning industries use natural gas with woodfuel as a supplemental fuel, to decrease the emissions of air pollutants resulting from woodfuel combustion, and/or to meet rapid swings in combustor load demand. On a smaller scale, some companies have site-specific combustion problems with woodfuel that can be corrected by cofiring with gas, such as greater boiler efficiency requirements, flame or temperature instability, and combustor maintenance. Natural gas competes with oil in all of these applications.

Supplemental Fuel. Many users of woodfuel do not produce enough waste wood to meet their energy needs, and they must therefore use a supplemental fuel. Natural gas is often used to make up this shortfall. Natural gas has a significant advantage in that it is available at a moment's notice through the pipeline, with no need for on-site storage.

Natural gas is used not only in woodfuel mixtures in multi-fuel boilers, but in boilers that burn only gas or oil as well. Many of these nonwood capable boilers still exist in the paper and lumber industries, but the trend has been to replace them with wood capable boilers as the gas boilers are retired.

Environmental. Uncontrolled woodfuel combustion in boilers results in over twice the total suspended particulate matter than even uncontrolled coal combustion (4.88 to 6.87 lbs/MMBtu heat input for woodfuel versus 2.54 lbs/MMBtu for coal). While there are no Federal environmental regulations (new source performance standards) specifically designed for wood-fired industrial boilers, many states regulate the particulate matter emissions of these boilers through state implementation plans required by the Clean Air Act. These emission standards, on average, range from 0.3 lbs/MMBtu to 0.4 lbs/MMBtu.[69]

About half of the wood-fired industrial boilers employ some form of a particulate matter control device. For 80 percent of these boilers, centrifugal collectors (cyclones, multiple cyclones, and dual mechanical collectors) are used. However, wood boiler particulate emissions with these collectors averaged above 0.5 lbs/MMBtu in EPA tests.[69]

Using a natural gas/woodfuel mixture would bring down these emissions even further, since natural gas combustion in boilers results in almost no particulate emissions (0.002 lbs/MMBtu). Thus, using a 50/50 gas/woodfuel mixture in a boiler with a mechanical collector would at least halve the particulate emissions for a wood-fired boiler emitting 0.6 lbs/MMBtu to a level of 0.3 lbs/MMBtu. In addition to the benefit from simply substituting a very low particulate fuel for one which has a very high particulate content, gas consumed in a "reburn" mode can achieve greater than one-for-one reductions in particulate emissions by incinerating some of the particulate matter produced by the woodfuel.

Ignition and Rapid Load Swings. The cold start-up of a boiler using woodfuels almost always requires the use of a fossil fuel (e.g., natural gas or a gas/woodfuel mixture) for ignition. In addition, using gas for preburn purposes helps the combustor achieve its optimal temperature in a shorter time than could be achieved by woodfuels alone. For example, air-dried wood requires anywhere from one minute to over 30 minutes for ignition, depending on the combustion chamber temperature.[70]

Gas mixtures also shorten the time required to meet rapid changes in the level of heat output. Woodfuels cannot respond quickly to sudden load swings. These changes in temperature or steam output can be better achieved through controlling the amount of gas used in the fuel mixture.[71] Boiler manufacturers have developed a fuel feeding system for some wood-fired systems which improves the response to load swings, but it also increases the capital costs which can offset the fuel cost savings achieved by backing out gas.[72]

Efficiency Gain. Woodfuel has a lower heating value and boiler efficiency when compared to fossil fuels (see Table 3.6.). Because woodfuel has a high moisture content, its heating value averages about 9,000 Btu/lb and its efficiency in boilers is in the range of 60 to 70 percent. While predrying the woodfuel increases its efficiency in boilers, this

Table 3.6.   A Comparison of Natural Gas and Woodfuels

	Natural Gas	Woodfuels
Heating Value (Btu/Lb)	22,000[1]	9,000[2]
Boiler Efficiency (%)	80-87[1]	60-70[3]
Boiler Combustion Emissions (Lbs/MMBtu)		
Particulate Matter	0.01[4]	4.88-6.87[3]
Sulfur Dioxide	negligible[4]	0.02[3]
Nitrogen Oxides	0.23[4]	0.25[3]
1983 Industrial Consumption (10^9 Btu)	6,822[5]	1,690[6]

[1]ICF Inc. for the Congressional Budget Office, *Economic Considerations in Industrial Boiler Fuel Choice* (Washington, D.C.: June 1978).

[2]David A. Tillman, *Wood as an Energy Resource Academic Press* (New York, N.Y.: 1978).

[3]Environmental Protection Agency, *Nonfossil Fuel Fired Industrial Boilers--Background Information* (Research Triangle Park, N.C.: March 1982).

[4]Environmental Protection Agency, *Federal Register*, Vol. 49, No. 119 (Washington, D.C.: June 19, 1984) P. 25106.

[5]American Gas Association, *Gas Facts, 1984 Data* (Arlington, Va.: 1985).

[6]Energy Information Administration, Department of Energy, *Estimates of U.S. Wood Energy Consumption 1980-1983* (Washington, D.C.: November 1984).

drying process consumes a significant amount of the energy gained. In addition, the drying process results in low grade oxidation and some pollution.[69]

Natural gas has a heating value of about 22,000 Btu/lb and a boiler efficiency of about 80 to 87 percent.[73]   By using a mixture of woodfuel and natural gas, more useful heat can be produced from a boiler than could be obtained by just using woodfuel.

Flame/Temperature Stabilization. Woodfuel combustion does not usually produce a precise flame or temperature, which makes it difficult to control the exact output of a boiler or process heater. While some industrial processes can tolerate moderate heat fluctuations, others cannot. The amount of energy provided by woodfuel cannot be precisely measured because woodfuel's energy content will vary, even within the specific woodfuel type. Rapid variations in woodfuel moisture content create difficulties in controlling combustion air. Finally, woodfuel burns erratically, with some pieces burning quickly at high temperatures and others slowly at lower temperatures. This temperature variation cannot be adequately controlled by adjusting the fuel input because of the time required for woodfuel ignition. Additionally, boilers cannot support stable combustion if the woodfuel's moisture content is too high.[69]

The combustion of natural gas is easily controlled. Ignition is instantaneous and energy input can be controlled accurately. For these reasons natural gas is often used with woodfuel for stabilizing the temperature output--the amount of gas input is varied to offset the combustor's temperature variations.

Reliability of Operation. Since natural gas contains virtually no impurities, it is a very clean burning fuel. Woodfuels, on the other hand, contain many impurities that either escape from the combustor through stack emissions and ash disposal, or remain in the combustor. The impurities that remain can cause fouling or slagging of the combustor, which causes downtime for cleaning. Gas/woodfuel mixtures reduce the amount of impurities, both as a result of mixing with a clean fuel and because of the higher combustion temperature, and thus contribute to a higher reliability of operation.

Industrial Power Production Options --
A Comparison of Natural Gas Boilers and Coal-Fired Fluidized Bed Boilers

Although the operating experience of fluidized bed boilers (FBCs) is limited, this coal option offers capital costs which are equal to or lower than coal units with scrubbers. In addition, FBCs can operate at a lower cost than some conventional coal boilers due to their ability to burn a lower grade of coal while meeting environmental restrictions.

The purpose of the analysis reported in this section is to examine the relative economics and environmental impacts of natural gas boilers versus coal-burning FBCs. The baseline case examines a typical industrial application requiring 87,000 pounds of steam output per hour (lbs/hr), or roughly 100 million British thermal units of fuel input per hour (MMBtu/hr). Alternate cases were run to examine the economic sensitivities of: differing fuel prices, capacity utilization rates, discount rates, unit sizes, and capital costs. In addition, the costs and environmental impacts of a conventional coal unit were included for comparative purposes.

For all ten of the cases examined in this analysis, natural gas use in industrial boilers was found to be more economical than coal use in FBCs. The FBC was found to be more economical than burning coal in a conventional boiler equipped with a scrubber. In addition, the emissions from a natural gas boiler were significantly less than those of the FBC.

Economic Results. Table 3.7 presents the levelized annual costs of providing 87,000 lbs/hr of steam from natural gas and coal units -- the cost of the gas option was found to be only 73 percent of burning coal in a FBC. While the fuel cost of the natural gas boiler is greater than the cost of coal for the FBC, the lower capital costs and operation and maintenance (O&M) costs of the gas boiler more than compensate for the fuel expense differential. Specifically:

- The levelized annual fuel cost of the gas boiler is twice that of the FBC -- $2.5 million per year versus $1.2 million -- since the FBC can burn a low grade coal with a relatively high sulfur content.

- The annual O&M expense of the natural gas boiler ($0.8 million) is only 41 percent that of the FBC ($2.0 million), due to the FBC's additional requirements for labor, limestone, waste disposal, and electricity.

- The levelized capital cost of the gas boiler is only 22 percent of the FBC's costs -- $0.4 million per year versus $1.9 million -- a result of the much higher up-front capital cost for FBC units.

Table 3.8 shows the base case and the sensitivity case results for the larger units (175,000 lbs/hr capacity) and also for higher and lower fuel price differentials.

- While the total levelized cost of the gas boiler is still less than that of the FBC for the larger unit, this advantage decreases as unit size, and thus the relative importance of fuel costs, increases.

- Increasing the fuel cost advantage of coal relative to gas from approximately $2.00 per MMBtu assumed in the base case to $3.00 per MMBtu results in a total levelized cost for the gas option which is 84 percent of the FBCs. Conversely, decreasing the fuel cost advantage to $1.25 per MMBtu increases the attractiveness of the gas option -- the total levelized cost is 64 percent of the FBC.

Table 3.7 also presents the costs of a comparably sized conventional coal boiler with scrubbers for comparison with the FBC system. Overall,

80

Table 3.7. "Typical" Levelized Annual Costs of Producing 40 Tons/Hr[1] of Steam for Industrial Processes (000 of 1985$)

	Natural Gas Boiler	Coal FBC	Coal Boiler w/ Scrubber
Fuel	2,468	1,221	1,238
O&M	807	1,987	2,190
Capital	407	1860	2,133
Total	3,682	5,068	5,561

[1]Approximately equal to a fuel input of 100 MMBtu/hr for the gas and coal boilers, and 106 MMBtu/hr for the coal FBC.

Sources: Natural gas and coal boilers: PEDCo-Environmental, Inc. for the U.S. Environmental Protection Agency, *Capital and Operating Costs for Industrial Boilers* (Research Triangle Park, N.C.: June 1979); Radian Corporation for the U.S. Environmental Protection Agency, *Costs of Sulfur Dioxide, Particulate Matter, and Nitrogen Oxide Controls on Fossil Fueled Industrial Boilers* (Research Triangle Park, N.C.: 1982).

FBC: Government Institutes Inc., *Evaluating the Fluidized Bed Combustion Option, Proceedings of the 2nd National Conference,* (Rockville, Md.: May 1984); Bibb and Associates Inc., *Test and Evaluation Period of 120,000 PPH Atmospheric Fluidized Bed Combustion Boiler with 3,500 kW Cogenerated Electric Power* Midwest Solvents Co. of Illinois and the State of Illinois Department of Energy and Natural REsources (Springfield, Ill.: March 1985); Phone conversations with various FBC manufacturers.

Fuel Costs -- National Coal Association, *Steam Electric Plant Factors 1984* (Washington, D.C.: 1984); American Gas Association, *A.G.A. -TERA Base Case 1985-I* (Arlington, Va.: February 1, 1985).

Major assumptions: Capacity utilization, 60%; Real cost of capital, 10%; Income tax rate, 50%.

the total costs are lower for the FBC process, mainly due to lower levelized annual O&M and capital costs.

While gas was shown to be more economical in all the cases examined, site-specific factors may result in some instances where coal FBC economics will prevail. Conversely, FBCs have certain operating characteristics which must be considered: limited operating experience; lower operating efficiencies; and, greater requirements for limestone, electricity, and waste disposal.

Table 3.8. **Summary Table of Sensitivity Case Total Levelized Annual Costs (000 of 1985$)**

Case	Natural Gas Boiler	FBC Coal Boiler	Conventional Coal Boilers
Base Case	3,682	5,068	5,562
80 Ton/Hr Unit	7,092	8,624	9,345
Higher Fuel Price Differential	4,040	4,837	N/A
Lower Fuel Price Differential	3,412	5,332	N/A
Higher Capacity Utilization	4,318	5,508	N/A
Lower Capacity Utilization	3,046	4,629	N/A
Higher Cost of Capital	3,734	5,482	N/A
Lower Cost of Capital	3,646	4,712	N/A
Lower FBC Capital Cost	3,682	4,510	N/A

Environmental Impact. Operating a natural gas boiler has much less of an impact on the environment than does operating a comparably-sized coal FBC boiler. Table 3.9 compares the annual air emissions from a natural gas boiler (employing a combustion technique that reduces NO_x emissions), a coal FBC boiler with a fabric filter for particulate matter (PM) control, and a conventional coal boiler with a scrubber and electrostatic precipitator.

The natural gas boiler emits only 11 percent of the particulate matter that the FBC emits. The FBC emits a much greater amount of this pollutant than a conventional coal boiler because of the increased amount of solid matter, such as unburned carbon, limestone, and gypsum, that escapes in the flue gas.

The natural gas boiler emits less than one percent of the sulfur dioxide that the FBC emits -- less than one ton of SO_2 for the gas boiler compared to 334 tons for the FBC. The FBC was assumed to control 70 percent of the SO_2 emissions, less than the 90 percent reduction achieved by the scrubber on a conventional coal boiler (removal efficiency for the FBC can be increased, but this requires a greater amount of purchased limestone).

The NO_x emitted from a gas boiler is only 24 percent that of the FBC. The FBC's NO_x emission numbers are 22 percent less than the uncontrolled spreader stoker coal boiler used here (the reduction would be greater when compared to a pulverized coal unit) due to the FBC producing less thermal NO_x with its lower combustion temperatures.

The gas boiler emits only six percent of the carbon monoxide that the FBC unit emits. The FBC unit emits a high amount of CO, even when compared to the conventional coal unit, because of the chemical reactions

Table 3.9. Annual[1] Emissions from Industrial Boiler Operations (Tons/Year)

	Particulates	Sulfur Dioxide	Nitrogen Oxides	Carbon Monoxides
Natural Gas[2]	0.63	0.16	30.60	8.94
Coal FBC[3]	5.85	334.28	128.14	144.80
Coal Boiler with Scrubber[2]	0.70	155.50	164.20	58.71

[1]Based on an output of 87,000 lbs/hr of steam operated at a 60 percent capacity utilization rate.

[2]Based on emission factors from U.S. Environmental Protection Agency, *Compilation of Air Pollutant Factors, Third Edition: Supplement No. 13* (Research Triangle Park, N.C.: August, 1982).

[3]Based on Bibb and Associates Inc.,*Test and Evaluation Period of 120,000 PPH Atmospheric Fluidized Bed Combustion Boiler with 3,500 kW Cogenerated Electric Power*, Midwest Solvents Co. of Illinois and the State of Illinois Department of Energy and Natural Resources (Springfield, Ill.: March 1985); Robert Gamble and William McCloy,*Update of the Georgetown Experience: Fluidized Bed Combustion* Foster Wheeler Energy Corp (Livingston, N.J.: April 1980); M.E. McDowell, "Case History: Choosing FBC over Conventional Boilers at Quaker State's Newell, West Virginia Refinery" *Evaluating the Fluidized Bed Combustion Option, Proceedings of the 2nd National Conference* Government Institutes Inc. (Rockville, Md.: May 1984); Bob Schwieger, "Fluidized-Bed Boilers Achieve Commercial Status Worldwide" *Power* (February 1985).

that occur between the limestone and sulfur in the combustion chamber which increase the amount of CO normally emitted during coal combustion.

Natural Gas and Climate Change: The Greenhouse Effect

The term "greenhouse effect" refers to the possibility that changes in the composition of the Earth's atmosphere due to human activity could result in a rise in average temperatures and severe weather disturbances. The greenhouse theory is controversial in the scientific community -- with the rate of possible warming and even the existence of a greenhouse problem subject to dispute.

Initially, concern about the greenhouse effect was focused primarily on an increasing atmospheric concentration of carbon dioxide resulting from fossil fuel combustion and deforestation. In recent years, however, research has also identified many other emitted gases that may contribute to Earth-warming by effectively trapping solar energy.[74]

Because the gases other than carbon dioxide are at relatively low atmospheric concentrations, they account for only about 10 percent of atmospheric radiative trapping (i.e., capture of solar energy) today. Water vapor, clouds and carbon dioxide account for 90 percent.[75] Some scientists have speculated that the buildup of all of the additional trace gases taken together could be about as important as the buildup of additional carbon dioxide. The carbon dioxide buildup is thought to be roughly one-half (possibly more) due to fossil fuel combustion and one-half due to deforestation.

Besides carbon dioxide, three of the trace gases -- nitrous oxide, chlorofluorocarbons and methane -- are the primary focus of greenhouse study. Other nitrogen oxides, non-methane hydrocarbons and carbon monoxide, while not greenhouse gases per se, are also of interest because they may affect the concentration of greenhouse gases by altering atmospheric chemistry.

Whether considering just carbon dioxide emissions or the full spectrum of air emissions, use of natural gas in lieu of other fossil fuels provides a net benefit from a greenhouse perspective. Encouraging use of natural gas in lieu of other fossil fuels offers a highly cost-effective element in a greenhouse response strategy.

Using natural gas instead of coal in boilers reduces carbon dioxide emissions by approximately 50 percent per unit of energy output.

Using natural gas instead of coal to generate electricity reduces carbon dioxide emissions per unit of electricity by approximately 65 percent.

Using natural gas instead of gasoline or diesel fuel in motor vehicles reduces carbon dioxide emissions per mile by approximately 30 percent.

In both stationary and vehicular applications, using natural gas in lieu of other fossil fuels also dramatically reduces emissions of a range of other pollutants which are either themselves important greenhouse gases or which result in increased concentrations of greenhouse gases, such as ozone, through their role in atmospheric chemistry.

Producing and using natural gas reduces methane emissions from landfills and coal seams, as well as seepage from natural gas reservoirs. These reductions appear to be large, but are not yet well quantified, so that it is not clear whether natural gas industry operations reduce or increase net methane emissions.

- Total atmospheric methane emissions resulting from *worldwide* natural gas industry operations appear to account for less than 5 percent of total worldwide methane emissions, and less than 1 percent of the total greenhouse effect.

- *Known* venting of methane from US natural gas industry operations (0.2 percent of production) totals about half of *known* methane venting from US coal mines (47 Bcf vs 90 Bcf annually), and as little as 5 percent of estimated methane emissions from landfills.

- Best available *estimates* indicate that *total* atmospheric methane emissions resulting from US natural gas industry operations (0.5 percent of production) account for 0.4 percent of total worldwide methane emissions. These methane emissions, necessary to achieving the significant greenhouse benefits outlined above, contribute about 0.1 percent (one tenth to three tenths of a percent) of the total worldwide greenhouse effect.

Concern has recently been voiced regarding methane emissions from vehicles fueled with natural gas, because methane is the primary hydrocarbon emitted from these vehicles. When methane emissions are factored in, however, the net effect of fueling a vehicle with natural gas instead of gasoline or diesel fuel is still to reduce the greenhouse impact of the vehicle by at least 25 percent.

- Even in a worst case analysis, giving the natural gas vehicle no credit for emissions reductions other than carbon dioxide and charging the vehicle for all methane losses in natural gas extraction and delivery, vehicular use of natural gas still clearly has a net greenhouse effect benefit.

Worldwide Methane Emissions

Methane is often referred to as one of the fastest growing greenhouse gases. The growth rate of methane is considered to be 1 percent per year, roughly two-to-three times the percentage growth rate of carbon dioxide. The total concentration of methane in the atmosphere, however, is only about 1/200 that of carbon dioxide, and it has a relatively short average atmospheric life of only 11 years. Consequently, the absolute growth rate of the number of molecules of methane in the atmosphere is only 1/74 that of carbon dioxide.

Many analysts consider a molecule of methane to be more effective as a greenhouse gas than a molecule of carbon dioxide, although the degree of effectiveness is in dispute. Assuming that methane actually is more

85

effective as a greenhouse gas than carbon dioxide, and considering a number of other factors, analysts conclude that methane accounts for approximately 15-20 percent of the future greenhouse problem.

Although the methane concentration in the atmosphere appears to have been growing for several centuries, there is considerable uncertainty as to why. The observed increases appear to be partially the result of decreases in the rate of destruction of atmospheric methane and partially the result of increased methane emissions. The decrease in the destruction rate is thought to result from increasing pollution which causes depletion of the atmospheric hydroxyl radical, the primary "sink" (i.e., remover) for atmospheric methane. Although the relative importance of this factor is not well established, it has been hypothesized that growing methane emissions are considerably more significant.[76,77,78,79]

The sources of methane emissions also are subject to considerable uncertainty, and the categorization of the data is often confusing.[80,81] It is currently estimated that increased human activities account for up to 60 percent of total methane emissions.[82,83,84,85] This figure includes biogenic sources such as increased termite methane emissions from disturbed forests, increased rice paddy emissions from multiple cropping, and ruminant animals (e.g., cattle and sheep), as well as biomass burning and coal and natural gas extraction and use.

Unfortunately, many analysts lump biomass burning and coal mining together with natural gas industry activities when estimating methane emissions. These three together, according to some current estimates, account for as much as 30-40 percent of total methane emissions.[86] Some other estimates are much lower.[87,88,89]

Uncertainties increase even further when one attempts to separate coal and natural gas industry methane emissions from biomass burning emissions. Clearly, however, it can be safely estimated that the world natural gas industry accounts for no more than 5-8 percent of world methane emissions, and no more than 1-1.5 percent of the total greenhouse effect, even if methane proves to be a highly effective greenhouse molecule.[90,91,92,93,94,95,96]

These figures ignore the fact that the world natural gas industry is also a means of reducing methane emissions, as when coal mine or landfill methane is captured and utilized. It has been estimated that US landfills alone might generate as much as 4 percent of worldwide methane emissions.[97]

Concern is sometimes voiced regarding methane emissions from vehicles fueled with natural gas, because methane is the primary hydrocarbon emission from these vehicles. However, a car emits approximately 1,000 times as much carbon dioxide (weight basis) as it does hydrocarbons. Given that natural gas use reduces carbon dioxide emissions by 30 percent, the effect is that, for every molecule of methane emitted, the natural gas vehicle reduces carbon dioxide emissions by 115 molecules. Even, for the sake of argument, allowing for the disputed 30-fold effectiveness of methane as a greenhouse gas, the net effect is to reduce the greenhouse impact of the vehicle by roughly 25 percent compared to what it would have been with gasoline.[98]

The Viability of Natural Gas Vehicles as a Means of Improving Air Quality

US EPA Estimates of Emission Reductions from NGVs. On January 28, 1988, the US Environmental Protection Agency issued a Technical Guidance Document, suitable for use by state and local governments in estimating the actual impact of various air quality improvement options. The document in question, entitled *Technical Report: Guidance on Estimating Motor Vehicle Emissions From the Use of Alternative Fuels and Fuel Blends*, estimates that NGVs can yield the following emission reductions:

- A 50% reduction in tailpipe emissions of carbon monoxide.

- A 40% reduction in tailpipe emissions of reactive hydrocarbons.

- A *100%* reduction in emissions of reactive hydrocarbons from filling stations and evaporation.

Other Estimates of Emission Reductions from NGVs. A.G.A.'s own estimates, based on gas industry experience, show that even more dramatic reductions are achievable for carbon monoxide and reactive hydrocarbons.

As compared to gasoline vehicles, natural gas vehicles offer major reductions in emissions of carbon monoxide and reactive hydrocarbons. When vehicles which use older gasoline technology are retrofitted to natural gas, nitrogen oxides can also be reduced significantly.

Other gaseous-fueled vehicles, such as propane vehicles, can also offer emission reductions. However, we will focus on natural gas vehicles because that is the subject we know best.

Today, nearly all natural gas vehicles on the road are conversions of existing vehicles that once relied solely on gasoline or diesel fuel. Experience with such vehicles indicates that, when proper procedures are used to convert older gasoline or diesel fuel vehicles to natural gas, such conversions can reduce emissions of carbon monoxide by up to 99%; can reduce reactive hydrocarbons by up to 92%; and can reduce emissions of nitrogen oxide by up to 65% (Table 3.10).

The actual emission reductions achieved will often be somewhat lower, depending on a large number of factors -- including the tuning of both engines; the age and design of both vehicles; and the condition and technology of both the gasoline emission controls and the natural gas conversion equipment. Despite these possible variations, however, the emission reductions will be significant so long as proper procedures and equipment are employed.

Of course, many of the gasoline vehicles on the road are not older vehicles with relatively unsophisticated technology for the control of gasoline emissions. Vehicles of more recent vintage are common.

Therefore, A.G.A. has been working to develop estimates of achievable emission reductions from shifts to natural gas by vehicles with new gasoline technology. It is now estimated that carbon monoxide could be reduced by up to 82% and reactive hydrocarbons could be reduced by up to 87% -- even for gasoline vehicles that are right off the assembly line, with new gasoline technology that has not suffered any degradation from exposure to road conditions.

Carbon monoxide and reactive hydrocarbons are the pollutants whose concentration levels pose the greatest challenge for many areas as they strive to meet Federal air quality standards. Therefore, the ability of NGVs to reduce these emissions is of special importance.

Nitrogen oxides -- a serious but secondary problem -- are a different proposition. Because new gasoline vehicles have made substantial headway in reducing gasoline-related nitrogen oxides, outperforming gasoline on this pollutant is a difficult challenge. However, experience indicates that, with proper procedures and the latest NGV technology, we can hold our own against gasoline on nitrogen oxides.

The evidence available also suggests that nitrogen oxides emissions, while a manageable problem for properly retrofitted natural gas vehicles, need be even less of a concern with newly manufactured engines. Such engines, because they are designed from the start for natural gas or "dual capability," can be engineered to prevent any nitrogen oxide problem before it begins.

In this regard, a newly released report on this subject by Professor E.J. Durbin of Princeton University holds great promise for the near future.[9] To quote from the study's key conclusions:

- This study shows from both an emissions and efficiency point of view, gaseous fuels, and in particular natural gas, are superior to liquid fuels in vehicular applications.

- Principal emission advantages of natural gas fuels come from the ability to more readily create homogenous mixtures of air and fuel,

Table 3.10. Estimates of Achievable Emission Reductions per Vehicle when Light Gasoline Vehicles are Retrofitted to Natural Gas[*]

Pollutant	Vehicles Using Older Gasoline Technology	Vehicles Using New Gasoline Technology
Carbon Monoxide	-99%	-82%
Reactive Hydrocarbons	-92%	-87%
Nitrogen Oxides	-65%	0

[*]Assumes use of proper procedures and equipment.

---

which permits the use of lean combustible mixtures, and from the low carbon hydrogen ratio of the fuel molecules.

- Most assessments of emissions with natural gas fuels are based on fuel systems that have not been optimized for natural gas. These systems do not exploit the full advantages of natural gas fuels.

- In an engine "designed for natural gas" it should be possible to achieve oxides of nitrogen (NO_x) emissions much lower than those emissions from gasoline engines.

- These low emissions should be attainable without catalytic converters to clean up the engine exhaust.

- Total hydrocarbons (HC) can be reduced 33 to 50 percent when using gaseous fuels. When the gaseous fuel is natural gas, environmentally reactive hydrocarbons can be reduced 90% as compared with gasoline.

- With lean mixture control systems, carbon monoxide (CO) can be effectively eliminated from the exhaust of natural gas fueled engines.

- Carbon dioxide emissions are reduced approximately one third as compared with gasoline.

- An essential element in achieving these results, with good driveability, is a control system that adjusts air/fuel ratio to fit power requirements. Such a control system, coincidentally, provides increased fuel economy.

Of course, while newly manufactured NGVs can be expected to outperform retrofitted NGVs in emission levels and other respects, retrofitted NGVs are still the key to any *immediate* improvement in pollution from carbon monoxide and reactive hydrocarbons. NGV retrofitting is the ony way to put large numbers of alternative-fuel vehicles on the road within the next few years.

Use of Natural Gas in Heavy Duty Engines. Newly manufactured heavy duty trucks and buses must soon meet tightened EPA standards for heavy duty engines. Newly manufactured natural gas engines are an extremely attractive compliance option.

In 1988, newly manufactured natural gas buses, leased to the City of New York by Brooklyn Union Gas Company, were tested by US EPA at the EPA facilities in Ann Arbor, Michigan. The results of these tests show that emissions from the buses met, or far surpassed, the EPA standards for model year 1991 gasoline-fueled heavy-duty engines.

For carbon monoxide, emissions were 10.6 grams per brake horsepower-hour (g/b-hp/hr), compared to a standard of 14.4.

For nitrogen oxide, emissions were 1.4 b/b-hp/hr, compared to a standard of 5.0.

For *reactive* hydrocabons, which are only about 15% of the total hydrocarbons emitted, the bus emissions were extremely low. Even for total hydrocarbons, most of which do not appear to react chemically in the lower atmosphere, the bus emissions were 1.2 g/b-hp/hr. The standard is 1.3.

An Efficiency and Environmental
Comparison of Home Heating and Cooling Systems

In order to better understand which of the principal space conditioning (both heating and cooling) alternatives most closely align with energy conservation and environmental concerns, this section estimates the energy efficiencies, and environmental pollution of three residential systems.

The three space conditioning systems considered for a new residence are: (1) gas heating and electric air conditioning; (2) oil heating and electric air conditioning; and (3) an all-electric heat pump system for both heating and cooling. All three options considered involve new, currently widely available equipment.

The installation of gas heating equipment in new homes, coupled with central electric air conditioning for cooling utilizes less primary energy resources -- gas, oil and coal -- and results in less environmental pollution, than do the alternatives considered.

Energy Efficiency. Over the *total fuel cycle*, the average efficiency of utilization of non-renewable energy resources for residential space conditioning with a gas furnace plus a conventional electric air conditioner is superior to its alternatives. The total fuel cycle includes extraction, transport and processing of fuel, electricity generation where applicable, transmission, distribution and end-use.

- 72% total fuel cycle efficiency for recuperative gas condensing furnace plus conventional electric air conditioner (primary energy consumption of 89 MMBtu/yr per household).

- 66% total fuel cycle efficiency for an oil furnace plus conventional electric air conditioner (97 MMBtu/yr per household).

- 45% total fuel cycle efficiency for an electric heat pump (141 MMBtu/yr per household).

Environmental Quality. Over the total fuel cycle, NO_x, SO_x and suspended particulate emissions associated with residential space conditioning are 300% to 450% greater with an electric heat pump, and 120% to 220% greater with an oil furnace/electric air conditioner combination than with a gas furnace/conventional electric air conditioner combination. Solid wastes are 450% greater for the electric heat pump compared to the gas or oil furnace/conventional electric air conditioner combinations.

The seasonal performance factor for cooling via the heat pump, again as determined by Oak Ridge, is 180% regardless of region. Conventional electric air conditioning provides cooling in both the oil and gas heated homes with an efficiency of 200%. Cooling with the heat pump is roughly 10% less efficient than cooling with central air conditioning according to Gordian Associates in a study prepared for the US Department of Energy -- although some available units range up to 30% less efficient.

All space conditioning equipment chosen for this analysis is highly efficient and currently available. Over time, newer and even more efficient equipment will be marketed. For example, the gas-fired pulse combustion furnace with a heating efficienty of approximately 95% will be on the market later this year. Similarly, a more efficient electric heat pump with a two state compressor may soon be available. However, this equipment was not considered herein due to limited information regarding cost and performance.

The fuel inputs required for space heating and cooling, as determined by dividing the annual heating and cooling loads by the respective

91

efficiency factors, were multiplied by the projected prices of gas, oil and electricity for the period 1982 through 1995 to obtain annual space conditioning bills. Residential sector gas home heating oil and electric prices as projected by the TERA-II: Demand Marketplace Model were see *Consumer Impact of Indefinite Gas Price Escalator Clauses Under Alternative Decontrol Plans* (Arlington, Va., American Gas Association, November 6, 1981). Space conditioning bills for the 14 year period were annualized and discounted at a 10% real rate. The annualized cost derived indicates the yearly cost for heating and cooling over the period in today's dollars assuming a 10% real discount rate (i.e., above inflation).

A range of gas prices was considered due to uncertainty regarding the impact of "indefinite price escalator" clauses in gas contracts on gas decontrolled post-1985 as mandated by the Natural Gas Policy Act of 1978. The lower gas price projection is based on the assumption that escalators will be defused so that new gas prices in 1985 will not exceed the competitive market level. The high end of the range assumes gas decontrolled in 1985 will be temporarily (three years) forced above a market clearing level by escalators.

The cost of heating equipment was based on units with an output capacity of roughly 80,000 Btu's per hour, while cooling equipment costs were based on a 35,000 Btu per hour capacity. Costs were not varied by region, as this configuration would satisfy many regions of the country. In addition, in very warm regions where heating systems could be smaller and less costly, cooling systems would be larger, offsetting this advantage; the converse would be true in cooler regions.

Primary Resource Efficiency and
Environmental Considerations

Each of the space conditioning systems considered would require the extraction, delivery and use of some form of energy. Energy is used or lost from the point of production to the point of consumption (processing losses, transportation energy required, etc.), and the total energy delivered into the home in the form of heated or cooled air divided by the primary energy produced is referred to as the primary resource efficiency. The primary resource efficiency of the gas/electric system is 72% versus 66% and 45% for the oil/electric and heat pump systems, respectively. Thus, a home using gas heating and electric cooling equipment would require only 63% of the primary energy required for the heat pump system and 92% of that required for the oil/electric system. (See Tables 3.11 and 3.12 for detail and sources.)

The environmental pollution attributable to each of the three space conditioning systems, from the point of production to end-use, is also

**Table 3.11. Estimated Average Primary Energy Requirements and
Total Energy Cycle Environmental Pollutants of
Three Space Conditioning Alternatives**

	Gas Heat/ Electric Air Conditioning	Oil Heating/ Electric Air Conditioning	Electric/ Electric Heat Pump
Primary Energy Required per Household (MMBtu/Yr.)[1,2]	89	97	141
Environmental Pollutants (Thousand Tons/ End-Use Quad)[3]			
NO$_x$	180	210	550
SO$_x$	100	220	460
TSP	50	70	180
Solid Wastes	9,500	9,500	43,000

Sources: Environmental Pollutants -- *Environmental Impacts, Efficiency, and Cost of Energy Supply and End Use*, Volume I, Columbia, Md., Hittman Associates, January 1975; and *A Western Regional Energy Development Study*, Austin, Texas, Radian Corporation, August 1975.

Table 3.12. Total System Primary Resource Efficiencies for
Residential Space Conditioning (Percent)

	Natural Gas	Oil	Fossil Fuel Generated Electricity		
			Gas-Fired	Oil-Fired	Coal-Fired
Delivery Efficiency[1]					
Extraction	99.3	99.6	99.3	99.6	98.8
Processing	96.7	87.8	96.7	87.8	92.0
Transportation	96.6	98.0	96.6	98.0	98.1
Conversion	N/A	N/A	31.1	31.9	32.6
Distribution	97.1	98.0	91.6	91.6	91.6
Total	90.1	84.0	26.4	25.0	26.6
End-Use Efficiency[1]					
Heating[3,4,5]	87.0	84.0		170.0	
Cooling[5,6]	200.0	200.0		180.0	
Total[2]	116.0	113.8		172.6	
Total System Efficiency	71.8	65.9		45.4	

[1]Efficiency refers to the energy used or lost along any step of the system trajectory (processing transportation, etc.); thus, efficiency may be calculated as the energy output of a particular step divided by the energy input -- thereby indicating the amount of energy used or lost in that step.

[2]Weighted by heating and cooling loads (74% and 26%, respectively) in Philadelphia, Pa., representative of a "national average" load.

[3]Recuperative gas condensing furnace; Federal Trade Commission AFUE.

[4]Blueray oil furnace, Federal Trade Commission AFUE.

[5]The heating season performance factor of an electric heat pump with a C.O.P. of 2.75 ranges from 1.52 (New England) to 2.09 (Pacific Region) with a regional average of 1.70 as weighted by number of households per region according to: Oak Ridge National Laboratory for the U.S. Department of Energy, *Performance and Economics of the ACES and Alternative Residential Heating and Air Conditioning Systems in 115 U.S. Cities*, (Oak Ridge, Tenn.: March 1981). Cooling efficiency, from the same source, is 180% invariant with respect to location.

[6]Conventional air conditioning assumed to be 10% more efficient than cooling with a heat pump. See: Gordian Associates for the U.S. Department of Energy, *Heat Pump Technology*, (Washington, D.C.: June 1978).

indicated on Table 3.11. Again, the gas/electric option is the most attractive. The environmental comparison assumes that the best available control technology is utilized. Consequently, the environmental impacts are understated by the amount of electricity that is generated in plants which do not use best available control technology.

Conclusion

Natural gas resources are abundant, widely distributed and highly underutilized. By all measures - air emissions, solid wastes, water pollution, and water consumption - natural gas is by far the most environmentally desirable fossil fuel. Compared to other energy alternatives natural gas systems are also highly capital and resource efficient, requiring significantly lower capital investment to produce, deliver and utilize a unit of energy. The emergence of highly attractive natural gas-fired electricity generation technologies have enhanced these advantages. Natural gas can and should play a major role in environmental protection in many nations.

REFERENCES AND NOTES

1. US Department of Energy, Office of Environmental Assessment, *National Environmental Impact Projection II*, Washington, D.C., March 1980, pp. 3-17 - 3-24.

2. American Gas Association, *Fact Book: Energy, the Environment and Natural Gas*, Arlington, Va., October 1983.

3. US Environmental Protection Agency, Office of Air and Radiation, *Compilation of Air Pollutant Emissions Factors*, Research Triangle Park, N.C., October 1986.

4. US Department of Energy, Office of Environmental Assessment, op. cit., p. 3-12.

5. American Gas Association, *Carbon Dioxide Emissions From Fossil Fuel Combustion and From Coal Gasification*, Arlington, Va., September 2, 1977; and Gushee, David E., *Carbon Dioxide Emissions for Methanol as a Vehicle Fuel*, Washington D.C., Congressional Research Service, January 4, 1988.

6. Penner, J.E., Connell, P.S., Wuebbles, D.J., and Covey, C.C., Lawrence Livermore Laboratory, *Climate Change and its Interactions with Air Chemistry: Perspectives and Research Needs*, Research Triangle Park, N.C., Atmospheric Sciences Research Laboratory, US Environmental Protection Agency, September 1988, pp. 7-8.

7. World Meteorological Organization, *Global Ozone Research and Monitoring Project Report No. 16, Atmospheric Ozone 1985 Volume 1*, Washington, D.C., National Aeronautics and Space Administration, 1985, pp. 84,86.

8. Pierotti, D. and Rasmussen, R.A., "Combustion As a Source of Nitrous Oxide in the Atmosphere," *Geophysical Research Letter*, Vol. 3, 1976, pp. 265-267; and Weiss, R.F. and Craig, H., "Production of Atmospheric Nitrous Oxide by Combustion," *Geophysical Research Letter*, Vol. 3, 1976, pp. 751- 753.

9. Durbin, Enoch J., *Understanding Emissions Levels From Vehicle Engines Fueled With Gaseous Fuels*, Princeton, N.J., Princeton University, February 1989, p. 1.

10. Baines, Thomas, US Environmental Protection Agency, *A Presentation Before the "Natural Gas and Clean Air: An Alliance for America's Future" Conference*, Washington, D.C., April 19, 1988.

11. Penner, et al., op. cit., pp. 7-8.

12. Emission Control Technology Division, Office of Mobile Sources, Office of Air and Radiation, US Environmental Protection Agency, *Guidance on Estimating Motor Vehicle Emission Reductions From the Use of Alternative Fuels and Fuel Blends*, Washington, D.C., January 29, 1988.

13. Durbin, op. cit.

14. Guidance on Estimating Motor Vehicle Emissions Reductions, op. cit.

15. American Gas Association, *Written Comments of the American Gas Association Regarding the Viability of Natural Gas Vehicles as a Means of Improving Air Quality*, San Fransisco, Calif., Before the State of California Advisory Board on Air Quality and Fuels, February 2, 1989, p. 11.

16. Durbin, op. cit.

17. Gold, T., "Ancient Carbon Sources of Atmospheric Methane," *Nature*, Vol. 335, No. 6189, September 29, 1988, p. 404.

18. Office of Fossil Energy, US Department of Energy, *Methane Recovery From Coalbeds: A Potential Energy Source*, Morgantown, W.V., October 1983, p. V.

19. Mause, Philip J., et al., *Recovering Gas From Landfills: Resource Potential and Institutional Barriers*, Washington, D.C., Kadison, Pfalzer, Woodard, Quinn & Rossi for the American Gas Association, March 1980.

20. "Trace Gas Scenarios," *UNEP Economic Workshop on the Ozone Layers*, May 1986.

21. Ehhalt, D.H., "The Atmospheric Cycle of Methane," *Tellus*, Vol, 26, 1974, pp. 58-70.

22. Ehhalt, D.H., and Schmidt, U., "Sources and Sinks of Atmospheric Methane," *Pure and Applied Geophysics*, Vol. 116, 1978, pp. 452-464.

23. US Energy Information Administration, *Natural Gas Monthly*, Washington, D.C., December 1988, p. 7.

24. Ibid., p. 6.

25. US Environmental Protection Agency, Research and Development - Energy Minerals and Industry, *Energy/Environmental Fact Book*, Washington, D.C., March 1978, p. 24.

26. US Department of Energy, Office of Environmental Assessment, *National Environmental Impact Projection II*, Washington, D.C., March 1983, pp. 4-7.

27. General Electric, GER 3401, *STAG Combined Cycle Product Line and Performance Characteristics*, Schnectady, N.Y., 1984, p. 568316A.

28. American Gas Association, *Comparison of Capital Costs to Consumers of Alternative New Domestic Energy Supplies*, Arlington, Va., December 14, 1983.

29. American Gas Association, *Natural Gas-Fueled Combined Cycle Electric Power Generation -- An Attractive Option*, Arlington, Va., May 10, 1985.

30. American Gas Association, *An Economic and Environmental Comparison of Natural Gas and Coal Use for Large-Scale Industrial Cogeneration*, Arlington, Va., October 26, 1984.

31. Hay, Nelson E., Editor, *Guide to Natural Gas Cogeneration*, Lilburn, Ga., The Fairmont Press, Inc., 1988.

32. American Gas Association, *Repowering With Natural Gas -- An Electricity Generating Option*, Arlington, Va., October 17, 1986.

33. Fay, James A., Golomb, Dan S., and Zachariades, Savvakis C., *Feasibility and Cost of Converting Oil- and Coal-Fired Utility Boilers to Intermittent Use of Natural Gas*, Cambridge, Mass., Massachusetts Institute of Technology, December 1986.

34. Fay, James A. and Golomb, Dan S., *Economics of Seasonal Gas Substitution in Coal- and Oil-Fired Power Plants*, Cambridge, Mass., Massachusetts Institute of Technology, December 1987.

35. Armiak, Michael J., Correspondence to McKinney, Chris, Wisconsin Department of Natural Resources, Detroit, Mich., ANR Pipeline Co., October 29, 1985.

36. American Gas Association, *Industrial Power Production Options -- A Comparison of Natural Gas Boilers and Coal-Fired Fluidized Bed Boilers*, Arlington, Va., October 18, 1985.

37. Booth, R.C., Breen, B.P., Gallaer, C.A. and Glickert, R.W., Energy Systems Associates, *Natural Gas/Pulverized Coal Cofiring Performance Testing at an Electric Utility Boiler*, Chicago, Ill., Gas Research Institute, July 1987.

38. Wilkinson, Paul L., *Seasonal Fuel Switching to Natural Gas -- A Logical Approach to Reduce Sulfur Dioxide and Other Air Emissions*, A Presentation to the Air Pollution Control Association, Detroit, Mich., June 16, 1985.

39. Ban, Stephen D., Berkau, Eugene E., and Pratapas, John M., *New Gas Combustion Technologies for Emission Control*, A Presentation to the Conference on Select Use, Washington, D.C., August 1985.

40. American Gas Association, *Ozone and Carbon Monoxide: The Role of Natural Gas in Attaining Clean Air Act Compliance: 1988 Update*, Arlington, Va., March 25, 1988.

41. American Gas Association, *Consumer Costs of Natural Gas, Fuel Oil and Electricity in New Homes*, Arlington, Va., March 7, 1986.

42. American Gas Association, *Cost Advantages of Natural Gas in Existing Homes*, Arlington, Va., April 4, 1986.

43. American Gas Association, *A Life-Cycle Cost Analysis of The Market Potential of Residential Gas Air Conditioning*, Arlington, Va., November 30, 1984.

44. American Gas Association, *Gas Cooling vs. Thermal Energy Storage: Peak-Shaving Options*, Arlington, Va., April 13, 1988.

45. American Gas Association, *1988 Commercial Cooling Fact Sheet and Market Assessment Summary*, Arlington, Va., November 4, 1988.

46. US Department of Energy, Energy Information Administration, *Cost and Quality of Fuels for Electric Utility Plants -- 1986*, Washington, D.C., July 1987, pp. 27-29.

47. National Acid Precipitation Assessment Program, Vol. II, *Emissions and Control*, Washington, D.C., 1987, pp. 2-6, 2-8.

48. Ibid., pp. 2-42, 2-51.

49. Ibid., pp. 2-42, 2-51.

50. Ibid., pp. 2-42, 2-54.

51. Breen, B.P., and Winberg, S.E., "Natural Gas Cofiring for Nitric Oxide and Sulfur Dioxide Control," (unpublished), p. 13.

52. *Carbon Dioxide Emissions From Fossil Fuel Combustion and From Coal Gasification*, op. cit., Arlington, Va., p. 5.

53. Maulbetsch, McElroy and Eskinazi, "Retrofit NO_x Control Options," as printed in *Journal of the Air Pollution Control Association*, Pittsburgh, Pa., November 1986, pp. 1296-1297.

54. US Department of Energy, Energy Information Administration, *Costs and Quality of Fuels for Electric Utility Plants -- 1986*, Washington, D.C., July 1987, pp. 33-37.

55. National Acid Precipitation Assessment Program, Vol. II, op. cit., p. 2-11.

56. US Department of Energy, Office of Fossil Energy, *America's Clean Coal Commitment*, Washington, D.C., February 1977, p. A-4.

57. National Acid Precipitation Assessment Program, Vol. II, op. cit., p. 2-43.

58. Kurzynske, Richard F., *Gas Technologies for Emissions Reduction and Operational Benefits*, printed in A.G.A./EEI *Conference Proceedings -- Efficient Electricity Generation With Natural Gas*, Arlington, Va., November 1987, p. XV1-5.

59. *Carbon Dioxide Emissions from Fossil Fuel Combustion and from Coal Gasification*, op. cit., p.5.

60. Heap, M.P., *Gas Reburn and Reburn With Sorbent Injection*, printed in A.G.A./GRI *Workshop Proceedings -- Gas Technologies for Coal-Fired Boiler Emissions Control And Operational Enhancement*, Knoxville, Tenn., June 1988, pp. 21-22.

61. For comparison of uncontrolled emission rates in utility boilers see: US Environmental Protection Agency, Office of Air and Radiation, *Compilation of Air Pollutant Emission Factors*, Vol. I, Supplement A, Research Triangle Park, N.C., pp. 1.4-2 and 1.1-2.

62. SO_2 and CO_2 emission rates for combined-cycle units based on uncontrolled emissions, see Ibid. NO_x emissions from gas turbines can range from roughly 10 to 100 parts per million, depending on the control technology employed. A mid-level range of 42 ppm, which could be achieved through steam or water injection, was assumed in this analysis. Much lower levels -- in the 10-20 ppm range -- are being required by EPA in many states. See: Larry L. Larsen, *Changing Emission Standards for Stationary Gas Turbines*, (presented in Charleston, S.C.: November 15, 1988, at the A.G.A./EEI conference on Efficient Electricity Generation With Natural Gas).

63. *Code of Federal Register*, 40 CFR Subpart D, Part 60.42 through 60.44.

64. US Environmental Protection Agency, *Compilation of Air Pollutant Emission Factors*, Research Triangle Park, N.C., August 1982, Supplement 13.

65. US Environmental Protection Agency, ibid., p. 3.3.1-2.

66. Radian Corp., Emissions from Stationary Gas Turbines, Research Triangle Park, N.C., November 1984, p. 4-78.

67. US Environmental Protection Agency, *Energy/Environment Fact Book*, Washington, D.C., March 1978, p.24.

68. General Electric, GEA-11387, op. cit., p. 568316A.

69. US Environmental Protection Agency, *Nonfossil Fuel Fired Industrial Boilers -- Background Information*, and *Fossil Fuel Fired Industrial Boilers -- Background Information*, Research Triangle Park, N.C., March 1982.

70. Tillman, David A., *Wood As An Energy Resource*, New York, N.Y., Academic Press, 1978.

71. Georgia Institute of Technology, *The Industrial Wood Energy Handbook*, New York, N.Y., Van Nostrand Reinhold Co., 1984.

72. Telephone conversation on April 10, 1986 with Dick Ellis, Energy Policy Coordinator of Union Camp in Wayne, N.J.

73. ICF Inc., *Economic Considerations in Industrial Boiler Fuel Choice*, Washington, D.C., US Congressional Budget Office, June 1978.

74. Penner, J.E., Connell, P.S., Wuebbles, D.J., and Covey, C.C., Lawrence Livermore Laboratory, *Climate Change and Its Interactions With Air Chemistry: Perspectives and Research Needs*, Research Triangle Park, N.C., Atmospheric Sciences Research Laboratory, US Environmental Protection Agency, September 1988, p. 8.

75. Ibid., p. 18.

76. Khalil, M.A.K. and Rasmussen, R.A., "Causes of Increasing Atmospheric Methane: Depletion of Hydroxyl Radicals and the Rise of Emissions," *Atmospheric Environment*, Vol. 19, 1985, pp. 397-407.

77. Levine, J.S., Rinsland, C.P. and Tennille, G.M., "The Photochemistry of Methane and Carbon Monoxide in the Troposphere in 1950 and 1985," *Nature*, Vol. 318, 1985, pp. 254-257.

78. Thompson, A.M. and Kavanaugh, M., "Tropospheric $CH_4/CO/NO_x$: The Next Fifty Years," *Effects of Changes in Stratospheric Ozone and Global Climate, Vol 2*, United Nations Environmental Program Report, 1986.

79. Bolle, N.J., Seiler, W. and Bolin B., "Other Greeenhouse Gases and Aerosols: Addressing Their Role for Atmospheric Radiative Transfer," in: Bolin, B., Doos, B.R., Warrick, B. and Jager, D. (eds.), *The Greenhouse Effect, Climate Change and Ecosystems*, New York, John Wiley & Sons, 1986.

80. World Meteorological Organization, op. cit., p. 98.

81. Penner, et al, op. cit., p. 11.

82. Penner, et al., op. cit., p. 12.

83. World Meteorological Organization, op. cit., pp. 92-100.

84. Seiler, W., Holzafel-Pschorn, A., Conrad, R., and Scharffe, D., "Methane Emissions From Rice Paddies," *Journal of Atmospheric Chemistry*, Vol. 1, 1984, pp. 000-268.

85. Seiler, W., Conrad, R., and Scharffe, D., "Field Studies of Other Emissions From Termite Nests Into the Atmosphere and Measurements of CH_4 Uptake by Tropical Soils," *Journal of Atmospheric Chemistry*, Vol. 1, 1984, pp. 171-187.

86. Penner, op. cit., p. 11-12.

87. World Meteorological Organization, op. cit., p. 93.

88. Ehhalt, D.H., "The Atmospheric Cycle of Methane," *Tellus*, Vol. 26, 1978, pp. 58-70.

89. Ehhalt, D.H., and Schmidt, U., "Sources and Sinks of Atmospheric Methane," *Pure and Applied Geophysics*, Vol. 116, 1978, pp. 452-464.

90. "Trace Gas Scenarios," *UNEP Economic Workshop on the Ozone Layers*, May 1986.

91. World Meteorological Organization, op. cit.

92. Ehhalt, and Ehhalt and Smith, op. cit.

93. Office of Fossil Energy, US Department of Energy, *Methane Recovery From Coalbeds: A Potential Energy Source*, Morgantown, W.V., October 1983, p. V.

94. Gold, T., "Ancient Carbon Sources of Atmospheric Methane," *Nature*, Vol. 335, No. 6189, September 29, 1988, p. 404.

95. Cicerone, R.J. and Oremland, R.S., *Biogeochemical Aspects of Atmospheric Methane*, Boulder, Colo., National Center for Atmospheric Research, and Menlo Park, Calif., US Geological Survey, August 17, 1988.

96. Assumes .0448 lb/ft^3 (*Handbook of Chemistry and Physics*, Boca Raton, Fla., CRC Press, Inc., 1980, p. B-140) @ .4554 Kg/lb (*Handbook of Engineering Conversions*, Tulsa, Okla., Combustion Engineering, Inc., 1977). World methane emissions from Penner, op. cit., p. 11. Methane's contribution to greenhouse effect 15-20 percent as cited previously.

97. Mause, Philip J., et al., *Recovering Gas From Landfills: Resource Potential and Institutional Barriers*, Washington, D.C., Kadison, Pflazer, Woodard, Quinn & Rossi for the American Gas Association, March 1980.

ACKNOWLEDGMENTS

The author/editor wishes to express his thanks to the authors and copyright holders for permission to draw heavily upon the works incorporated into this chapter. Significant portions of the material were contributed by Paul Wilkinson, Paul McArdle and Michael German. Copyrighted materials were quoted with permission from the American Gas Association. The draft was typed and proofread by Lisa Hamako.

4

Natural Gas and Development: The Policy Issues for Developing Countries

Sergio C. Trindade

Introduction

Traditionally natural gas has been the unwanted child of oil exploration, which is at the root of the policy issues and options affecting gas development programs. The gas business is structurally very different from the oil business. The majority of exploration activity that hits a non-associated gas reservoir is searching for oil. Such gas finds are often considered dry holes from the oil point of view. On the other hand, associated gas is an unavoidable co-product of petroleum production. Its development in practice is subordinate to the development of the associated oil field.

Natural Gas: The Leading World Energy Resource in the Twenty-First Century

Natural gas promises to become the leading energy source into the twenty-first century. The logistical analysis of Marchetti[1] shows gas penetrating global energy markets in competition with oil, coal, nuclear fissile and fusion energies to reach a market share peak in 2030. By then gas should supply 70 percent of the global energy market. The absolute peak of gas consumption in a larger global energy market is expected by 2060-2070 to be at the level of about 30 trillion m^3/year. The final cumulative demand for gas over its total life cycle is foreseen to be about 2500 trillion m^3; that is, the energy equivalent to five times that for oil over their respective lifetimes as energy resources.

The fact that known present reserves are only of the order of 100 trillion m^3 should not be of concern, according to Marchetti. After all, oil reserves in the US have always been about 14 years of current consumption.

In reference to Grossling,[2] the point is made that "the search for hydrocarbons is much more controlled by geopolitical context than by the probability of finding them." Consequently, there must be plenty of potential gas bearing formations that will allow matching total demand forecast.

Such a scenario of gas dominance in the energy market of the twenty-first century requires the development of substantial infrastructures, particularly for long range transportation. That includes longer and larger diameter pipelines connecting continents, LNG tankers and new yet unknown technologies for natural gas transportation.

This prospective natural gas dominance of global energy will influence world affairs in a manner that oil trade does today. The current situation in Europe with an infrastructure of gas pipelines supplied from many non European countries, and particularly from the Soviet Union, illustrates the point.

Natural Gas Growth Likely to Take Place Mostly in the Developing Countries

Irrespective of the actual market penetration of natural gas in global energy markets over the next 100 years, the absolute market size of gas is likely to be many times greater than its current level. The history of natural gas development has evolved since the nineteenth century. But it was in this century that the development of gas transportation technology, by pipeline and otherwise, led to a faster pace of market penetration. In 1988 total world gas consumption reached 1960 billions of cubic meters; the OECD member countries accounted for some 45 percent of that total. Eastern Europe, including the totality of the USSR, accounted for another 40 percent. The reserves of natural gas in developing countries are huge relative to current predictions of demand. Fish as quoted by Schramm[3] summarized the long-term gas supply and demand balance as follows: "...in most areas of the world very significant reserves and undiscovered resources will still exist in 2020. About 2/3 of the world's currently identified proved and additionally recoverable gas will still not have been produced by year 2020... The quantity of remaining recoverable gas in 2020 indicates that natural gas will continue to be a major fuel well into the next century."

The known proven natural gas reserves as of 1st January 1989 total 116.3 trillion m^3. Developing countries' reserves amount to 59.9 trillion m^3. The OECD countries hold some 15.7 trillion m^3. The USSR and the Eastern European countries hold about 43.1 trillion m^3.

The actual reserves of gas in the developing countries could be much larger. Most of the gas finds so far resulted from oil exploration. As many more developing countries continue or initiate the search for oil in the future, the list of these countries with significant gas reserves is likely to increase. Consequently, much of the future gas development is likely to take place in developing countries as their domestic markets evolve and as a select few of them find export markets for their gas.

Natural Gas Markets:
Conventional and Unconventional

With current technologies the markets for gas in substitution for other energy resources depend on the respective total system costs. In many respects gas is a less convenient fuel to handle than oil. That is reflected in the costs of gas infrastructure, and the need of large scale transportation system for export markets. But domestic markets may offer in some cases an alternative to gas development, particularly when field development and delivery costs are low compared with competing energy resources, usually oil or coal. In building up a gas market supplying a few relatively large consumers is essential. Thus, the conventional markets for gas usually begin with power plants and energy-intensive industrial plants as fuel and chemical feedstock. This development may allow later on the development of residential and commercial markets.

Besides these conventional gas markets, once a market base is established, unconventional uses could be developed for co-products such as LPG (liquefied petroleum gas), for new markets such as in co-generation schemes and as transportation fuel, including aviation fuel, and as raw material for liquid fuels such as methanol, synthetic gasoline and synthetic middle distillates. On the supply side there is some hope for unconventional gas resources, small and large, which do not conform with the traditional concepts of associated and non-associated gas resources.

Natural Gas:
The Environmentally Sound Fossil Fuel

As the world grows aware of the global warming threat derived from the greenhouse effect produced by the increase in the atmospheric concentration of radiative active gases, the conventional economics of energy resources is bound to change. It appears inevitable that the externalities of energy systems, such as their environmental effects, are going to be eventually internalized in future analyses. Consequently, energy systems in the future will be compared on the basis of total costs and benefits that include the so far neglected environmental costs and benefits.

The radiative active gases include carbon dioxide (the end product of all fossil fuel burning) and methane (the main component of natural gas).

The amount of carbon dioxide per unit energy resulting from fossil fuel burning depends on the ratio of carbon-to-hydrogen in the fuel. The smaller the ratio the smaller is the amount of carbon dioxide emitted into the atmosphere per unit of energy. Natural gas is thus the fossil fuel that emits the least amount of carbon dioxide per unit of energy. In future energy economic calculations that internalize environmental costs and benefits, natural gas should be the least penalized among fossil fuels for carbon dioxide emissions.

On the other hand, as methane itself is a radiative active gas 25 times as effective as carbon dioxide in contributing to the greenhouse effect, natural gas systems should be penalized for the inevitable gas leaks into the atmosphere. Thus it can be said that burning methane into carbon dioxide and water is benign to the environment. As natural gas markets expand both these features should be factored in the new economics that internalizes environmental costs and benefits of energy systems.

In some visionary concepts for natural gas utilization both the methane and carbon dioxide issues are taken care of. The heat of a high temperature nuclear reactor supports methanol production from natural gas. The excess carbon dioxide from this process is used for additional oil recovery.

Main Obstacles to Gas Development

There is no world gas market today as in the case of oil. The market is segmented and consists of groups of bilateral and multilateral trade relationships. The major gas importing markets currently are Western Europe, Japan, Eastern Europe and the US.

The future of natural gas lies in expanding energy markets, not in oil substitution in stagnant markets. Market penetration of gas in the future will be hampered by the expected costs of bringing gas to markets.[4] The classical argument, of course, excludes the internalization of environmental costs. Another barrier is the convenience of liquid fuels relative to natural gas with respect to transportation and storage. This reflects the physical properties of gas in comparison with liquid fuels. It remains to be seen if the argument of convenience still holds true when environmental costs are fully internalized for all competing fuels.

The main obstacles to gas trade and to gas development at large are thus: the large capital costs of gas supply systems including distribution; and the high level of risk associated with long lead times of investment and rigidities of pipeline and LNG (liquefied natural gas) systems that constitute the infrastructure required to bring gas to markets.

Furthermore, considerable uncertainty surrounds future developments of energy prices, alternative supplies of competing fuels, and energy demand.

Political factors in all countries will affect the actual economics of natural gas and thus have an impact on rates of market penetration both domestically and in international trade. The policy issues related to gas development will therefore vary from country to country.

Economically Based Development Strategies

Economic theory provides a framework for the analysis of gas utilization schemes thus helping to rank proposed project packages. As described by Julius[5] and Mashayekhi,[6] this framework consists basically of:

- Establishing the economic price of gas, that is, the economic cost plus a depletion premium representing the future cost of gas replacement when the resource considered is depleted;

- Establishing the gas netback values and comparing with the economic price of gas, and selecting those projects which pass the test;

- Ranking combinations of projects for which netback exceeds economic price, according with net present values.

The basic methodological problem of optimizing gas utilization lies in the fact that the economic price of gas depends on the set of selected projects. Obviously in practice gas pricing policies play a key role in the actual implementation of gas development strategies.

The application of such methodology to actual cases depends on the nature of the gas resource and its potential markets. There may be country situations where the gas market is segmented, that is, reserves in different regions and markets or different kinds of gas resource (e.g. associated and non-associated gas).

The policy issues of gas development are very much connected with the utilization strategies considered.

For gas-surplus situations the prospective demand/supply balance is such that the point of economic depletion is far into the future. Hence for practical purposes the economic price path for gas becomes the long-run marginal cost of gas production and delivery. Thus project sequence issues will not be too important and increases in the net present value of projects will come mostly from bringing projects on stream more rapidly.

109

Constraints include lack of infrastructure, availability of capital, and adequacy of gas composition for direct fuel use.

For gas-short situations the potential gas availability is not likely to exceed over time the potential demand. In such situations the economic price path for gas will follow that of the fuel displaced by gas on an energy equivalent basis. When, in gas-short situations, there is a high value of gas, project selection is limited to the domestic market. Netback values of gas for selected projects are calculated and compared with the economic value of gas. Next, trial packages of gas utilization projects are formulated for projects passing the netback test. Their respective net present values are then calculated and the packages can then be ranked.

The strategies for developing associated gas markets are hampered by the connection between gas and oil production. The operational flexibility of oil production may make the availability of gas uncertain or make gas handling facilities more extensive than those for non-associated gas. Consequently, the costs of gathering and conditioning associated gas per unit of production are frequently higher than the costs of non-associated gas in the same area. On the other hand associated gas has usually a higher energy content than non-associated gas.

However, the investment required to develop non-associated gas fields is too large to depend on uncertainty about gas prices. On the other hand development of non-associated gas provide for the involvement of non-oil interests and, therefore, may allow for gas policies that are more effectively integrated into the overall energy policy of a country.

The Policy Issues of Gas Development

Gas development policy should not be a sectoral policy of its own, but must be part of a comprehensive energy policy. It need not be necessarily part of the oil policy even in the case of associated gas, despite the practical difficulties of the latter proposal.

Gas development needs to be seen in the context of overall energy development. Gas could displace heating fuels, could fuel power generation, could replace liquid fuels in transportation, could be a source of liquid hydrocarbons, could be a petrochemical feedstock, and would find other uses of its own that future technologies will determine.

Gas development strategies should be based on comprehensive assessment of options. However, national strategic objectives, unforeseen higher value uses and uncertainties in gas netbacks should be all factored in. Sensitivity of projects to the future value of petroleum and economic rates of discount are key concerns.

Implementing a gas development strategy may require setting up new organizations and/or motivating existing organizations. The role of governments is key, particularly when high commercial risks or conflicts of interest among participating organizations are likely to occur.

Eventually, successful development of gas markets could, in the long term, lead to integration of segmented markets in a given country or region with supply, transmission, distribution and consumption in diverse geographic areas.

Of particular interest to developing countries is the development of remote or small gas reserves. As gas market potential evolves in developing countries, decision-makers will be confronted with policy issues and options regarding gas strategies to follow.

The institutional basis for gas development will be an important consideration. The organized energy sector in most developing countries consists of the national power utility, the national oil company and the oil transnational corporations. Ministries of energy, where they exist, have usually been established in the last decade or two. Often they lack adequate resources, human and otherwise, to counterbalance the weight of the energy companies in establishing *de facto* energy policies.

The formulation of gas development policies could benefit from consensus building dialogues among stakeholders in the gas economy. The latter are constituencies likely to be affected positively or negatively by gas development proposals. The stakeholders in the gas economy include oil and electrical utility companies, banks, potential gas consumers in industry, scientists and engineers, members of the organized public, and politicians, etc.

The dialogue process is also conducive to building endogenous capacity in the gas economy. Endogenous capacity is the ability of countries, organizations and individuals to make reasonable and relatively independent decisions. Endogenous capacity is a concept that evolved out of the 1979 United Nations conference on science and technology for development. It applies to all fields of development and is the ultimate measure of development. Thus, as in other development domains the implementation of a gas industry should be used as a means to build endogenous capacity.

The consensus building stakeholders' dialogue could produce, among other results, an agreement about priorities in gas development. Steady but resilient policies are essential for full utilization of the potential that might exist in many countries.

Endogenous capacity in the gas economy requires a certain minimum indigenous competence in engineering, and research and development. This would be particularly important over the long term in support of existing markets and to develop new markets for gas. It would also be essential for recipient countries to take full advantage of gas technology transfer.

Once the institutional basis is organized and the rounds of stakeholders' dialogues take place, conflicts can be resolved and sustainable gas development policies can be formulated and implemented.

The main policy issues that the stakeholders in the gas economy must address are:

- Role of gas in the overall energy economy;

- Gas resources to consider, large, small; associated, non-associated; conventional, non-conventional;

- Choice of markets, domestic and international;

- Financing of gas development;

- Pricing of gas.

The above policy issues are interrelated. Nevertheless, two clusters can be identified for purposes of analysis. One cluster contains the first three issues and is more of a planning nature. The other cluster includes the two remaining issues which relate more to implementation stages of gas development.

In practice rounds of stakeholders dialogues must be carefully prepared with as much involvement of nationals of the country concerned as possible from the very beginning. If adequately planned and executed this could be the initial step in making gas development into one of many processes of true social and economic development. This should be a process of maturing over time in gas technology and in the gas development domain. In other words this process of planning should be a process of endogenous capacity building in gas economy.

Developing countries which pursue endogenous capacity development strategies take full advantage of any new opportunity to increase their decision-making capability and control over their destinies. This applies to gas development as well, among the many development domains. In reality not too many developing countries follow an endogenous capacity strategy when developing their gas resources. They miss a unique opportunity of improving the quality and the sustainability of their development process while implementing gas development programs.

There are no prescriptions for the best institutional setup for gas development. The best setup is the one that serves a country according to its peculiarities and provides for sustainability and resilience over time. The stakeholders' dialogues should directly or indirectly address the issue

112

of the institutional basis for gas development. The consideration of the role of gas in the overall energy economy is naturally suited to the participatory decision-making process that characterizes the stakeholders' dialogues.

Data on the gas resources are essential for the discussion on the role of gas in the economy. Too often opportunities to use small or non-conventional gas resources are overlooked either for lack of data or for lack of open-minded concepts for gas development. The opportunities to bring associated gas to markets are frequently complicated by issues related to the fact that gas is a co-product of oil exploitation and often plays an important operational role in oil recovery. One early conclusion in the process of planning for gas development is the need for data on reserves of all kinds of gas. Another one is the requirement of a minimum cadre of trained staff in both technical and managerial disciplines to undertake effective interactions with foreign partners interested in the development of gas resources in a particular country.

The consideration of markets depend of course on the size of the reserves, the availability of long term export markets, the net-back value of gas in domestic and international markets, and inevitability on political factors. The rounds of dialogues among stakeholders in the gas economy constitute an ideal process to resolve possible conflicts in gas development planning and to establish a consensus that would place gas policy on firm and sustainable grounds for the future.

Over time gas development in neighboring countries could lead to integrated gas economies in subregional contexts such as ASEAN countries, South Asia, Southern South America, Central America, West Africa, East Africa, and Northern Africa.

The process of implementing gas development plans must resolve the important practical questions of project financing and gas pricing. In many situations the large capital costs of gas supply and distribution systems require external financing. This increases the debt burden of the country concerned, but the implementation of the project in most cases could generate the equivalent to the foreign exchange required to pay back the debt incurred. Timing is of the essence to minimize the financial burden of the gas project considered. Smaller gas projects for domestic markets in situations of high netback gas values offer opportunities not too often considered. The smaller capital outlays required could make them attractive in select cases to national investors.

As there is no market crude equivalent for natural gas pricing, there is no universally agreed formula for gas prices. There are, in practice, three main concepts in gas pricing in use today, namely pricing based on gas cost, gas value and the gas opportunity cost.

An approach is pricing based on the cost of gas developed from the costing of gas delivered via pipeline. The ultimate cost of gas reflects the investment in the transportation infrastructure, operating costs and a preset rate of return on investment. The situation is complicated in the case of associated gas due to the problems of cost allocation between oil and gas and the determination of the level of risk for gas exploration to be reflected in the rate of return on investment. Furthermore, if gas is in a competitive market its eventual price must make it competitive with alternative fuels or feedstocks. The pricing of non-associated gas in this case is less flexible for the lack of the oil co-product.

Gas value based pricing reflects the next best alternative to gas at the point of use. There is ample scope for interpretation and negotiation between sellers and buyers as to the meaning of the next best alternative. The gas supplier would be interested in higher value products, such as distillates and lighter fractions, as the next best alternative. The consumer on the other hand might consider that lower value fuels such as coal and heavy fuel oil constitute the next best alternative. The other issue with this pricing concept is the calculation of well head netback price, particularly when there are sources of supply located at widely different distances from the point(s) of consumption. When the next best alternatives are limited to oil products often a weighted composite price of light and heavy oil products is used.

Pricing gas on the basis of opportunity cost results in establishing the highest market value for gas depending on the actual markets available as for example, for export market pipeline gas, LNG or gas-based petrochemicals. The gas netback is obtained by subtracting from the market value of the product considered, the conversion, transportation and conditioning costs. Since there is no gas marker price of reference the opportunity cost pricing of gas netback may introduce over time a larger degree of uncertainty than the other pricing concepts.

Whatever the pricing concept, natural gas must cover all fixed and variable costs of development, production and operation plus the return on investment that will attract capital. Gas can be used as a vector for industrial development which requires that gas prices be competitive in the market place.

Government has a role in gas pricing that must be exercised carefully as the international experience shows that gas development can be crucially dependent on pricing policies and regulations. The stakeholders' dialogues could help achieve consensus in this critical area.

Conclusions

Natural gas, traditionally the unwanted child of oil exploration, is likely to become the leading world energy resource before the middle of the twenty-first century. Furthermore, the growth of natural gas markets is bound to occur predominantly in today's developing countries, many of which could use gas as a leverage in accelerating their development process. There is plenty of scope for exploration, exploitation and development of natural gas in the world over the next century. New technologies will play a significant role in bringing gas from all sources into markets where new uses are likely to arise in the long-term future.

Methane, the main constituent of natural gas, is a highly radiatively active gas. Methane in the atmosphere is 25 times more effective than carbon dioxide in terms of the green house effect. Consequently, burning methane to carbon dioxide and water is one way of combatting global warming. Furthermore, of all fossil fuels natural gas is the one that emits the least amount of carbon dioxide per unit of energy. This results from the elemental composition of methane with four atoms of hydrogen and only one of carbon. As concerns over climatic change increase, the environmental costs and benefits of natural gas development are likely to be internalized in economic calculations thus facilitating market penetration of gas. Natural gas is the environmentally sound fossil fuel.

The main obstacles to gas development in the immediate future relate to the large fixed investment required to build a supply and distribution infrastructure which with current technologies is rather inflexible in terms of market destinations, and takes some time and financial costs to implement. These risks are compounded with uncertainties about future energy prices, alternative supplies of competing fuels, and energy demand. Management issues, contractual arrangements, and financing and guarantees are areas that warrant further attention and discussion aiming at reducing risk and uncertainty in gas development.

Economic theory provides a framework for the analysis of gas utilization schemes. Nevertheless, gas development, as any natural resource development, can be heavily influenced by political considerations. Sustainable and resilient planning and implementation of gas development programs could benefit from rounds of dialogues among stakeholders in the gas economy. Such ongoing participatory decision-making processes are at the root of successful gas development initiatives. If a particular stakeholder, for example the gas supplier, has more than a fair stake instability is likely to plague gas development.

As an instrument for development at large, natural gas offers a unique opportunity, availed by very few developing countries, to build endogenous capacity. This is the capacity to make reasonable and independent decisions in gas development matters. This capacity enhances the quality and the sustainability of a development process.

In implementing gas development programs, the stakeholders' dialogues will help create or strengthen the institutional basis required. They will also facilitate addressing the main policy issues involved in planning and implementing a natural gas economy. These issues include the role of gas in the overall energy economy, the gas resources to consider, the choice of markets, the financing of gas projects and the pricing of gas.

Overall gas development in developing countries must be:

- Planned and implemented in the context of overall energy development, but must be flexible relative to guidelines in the national context when applied to speific situations over time;

- Based on a comprehensive economic assessment of options, but must take into account national strategic objecties as defined through the stakeholders' dialogue process, and be sensitive to uncertain factors such as new technologies, the value of competing resources, etc;

- Built upon new organizations or on motivated existing entities, in a framework where the role of government, aided by the stakeholders' dialogues, is key in the resolution of conflicts and supporting sensible risk taking.

REFERENCES

1. Lucas, N.J.D., et al. *Energy Policies in Asia - A Comparative Study.* McGraw Hill Singapore, 1987. 480 pp.

2. Grossling, P.F. "Window on Oil: A Survey of World Petroleum Resources." *Financial Times*, London, 1986.

3. Schramm, Gunther. "The Changing World of Natural Gas Utilization." *Natural Resources Journal*, Vol. 24 (April 1984), pp. 405-36.

4. Mabro, Robert, editor. *Natural Gas - an International Perspective.* Oxford University Press, Oxford, 1986 155p.

5. Julius, DeAnne. *Natural Gas Utilization Studies: Methodology and Application.* World Bank, Washington, D.C., 1984.

6. Mashayekhi, A. "The Economics of Natural Gas Development." *Natural Gas: Its Role and Potential in Economic Development.* Westview Press, Boulder, Colo., 1990.

7. Trindade, Sergio C. *Isolated Domestic Energy Markets: The Case of Gas*, World Energy Conference Monograph, London, 1986. 15 pp.

8. Marchetti, Cesare. "World Status Report; Natural Gas." *Energy Economist*, No.95, September 1989, pp.15-19.

9. Marchetti, Cesare. *The Future of Natural Gas - a Darwinian Analysis.* International Institute for Applied Systems Analysis, Laxenburg. 1986. 7 pp., plus 19 figs.

10. World Bank. "Summary Proceedings of the Round Table of Executives of Oil Companies, Gas Companies and Senior Representatives of Developing Countries on Gas Development in Developing Countries," Energy Department, Paris, 25-26 March 1985, 12 pp.

11. Munasinghe, Mohan. "Energy Strategies for Oil Importing Developing Countries." *Natural Resources Journal*, Vol. 24, April 1984, pp. 351-368.

5

Power Generation with Natural Gas-Fired Gas Turbines

Robert H. Williams
Eric D. Larson

Introduction

A revolution is underway in electricity generating technology that may soon radically transform the power industry in both industrial and developing countries. This revolution involves not an exotic new technology, but rather an upgrading of the familiar but little-used gas turbine, the neglected step-sister of the steam turbine in power generation.

Though the gas turbine has always offered a low unit capital cost, it has traditionally been restricted in utility applications to peaking service -- for power plants that operate only during the short periods when the demand for electricity reaches its peak values. This is due in part to its low efficiency and in part to public policies that have constrained the use of natural gas for power generation. Its peripheral role in power generation also fostered less-than-rigorous maintenance practices, helping to give the gas turbine a reputation of poor reliability among utility planners.

Although in many parts of the world its use for cogeneration has been inhibited by institutional constraints, the simple cycle gas turbine's high thermodynamic efficiency for combined heat and power production as well as its low capital cost have long made it an economically attractive option for cogeneration applications. Nevertheless, the simple cycle gas turbine has been used mainly in cogeneration applications characterized by steady steam loads due to its poor part-load performance.

Innovations in technology, however, are making gas turbines competitive in cogeneration markets characterized by variable heat loads and in central-station applications with conventional baseload and load-following technologies. These technological innovations, a bullish outlook for natural gas supplies in many parts of the world, and tightening environmental constraints on power production are all important factors that point to future widespread use of natural gas-fired gas turbines for power generation.

It is useful to distinguish between heavy-duty industrial turbines (designed specifically for stationary applications) and aeroderivative

119

turbines (derived from jet engines). The heavy-duty industrial turbine is quite familiar in the power sector and is the technology of choice for the gas turbine/steam turbine combined cycle systems beginning to be widely used in various parts of the world. The various configurations of aeroderivative turbines of interest -- e.g., the steam-injected gas turbine, the intercooled steam-injected gas turbine, the intercooled steam-injected gas turbine with reheat, and the intercooled steam-injected gas turbine with reheat and chemical recuperation -- are much less well-known. This chapter will focus on aeroderivative turbines, because their various cycle modifications represent relatively new and unfamiliar developments and because this class of technologies offers advantages in some important markets. Both types of turbines have major roles to play in power generation; hence in the decades ahead a more balanced mix of heavy-duty industrial and aeroderivative turbines than is anticipated by most power planners will probably evolve in power markets.

The Policy Context

The electric power industry needs a technological revolution, since business-as-usual is becoming increasingly untenable.

Public concerns about nuclear power risks and the environmental problems posed by fossil fuel power plants have made electric utility planning more and more difficult. The Chernobyl accident has led to a considerable stiffening of public opposition to nuclear power and to sharp limitations on nuclear power programs in various parts of the world. The large quantities of SO_2 and NO_x emitted by existing fossil fuel plants, especially coal-fired plants, are among the most significant pollutant emissions leading to acid deposition, which has become a serious international problem because of the transnational air transport of these pollutants. The need to control acid deposition is leading to costly proposals to reduce emissions from existing power plants in some areas. Global warming from the greenhouse effect, associated with the buildup of CO_2 in the atmosphere, has also become a major environmental concern.

The electric utility industry has also been experiencing rising electricity costs -- a marked departure from the long-term historical trend, as illustrated by the situation in the US. Up until 1970 the outlook for electric power was bright in the US. Between 1950 and 1970, the real cost of fossil-fueled steam plants fell by more than a factor of two (Figure 5.1). Cost-cutting innovations in the US power industry persistently led to a reduction in the average real price of electricity by about 25% for each doubling of cumulative electricity production throughout the period 1926 to 1970 (Figure 5.2). If the trend up to 1970 had persisted to the present,

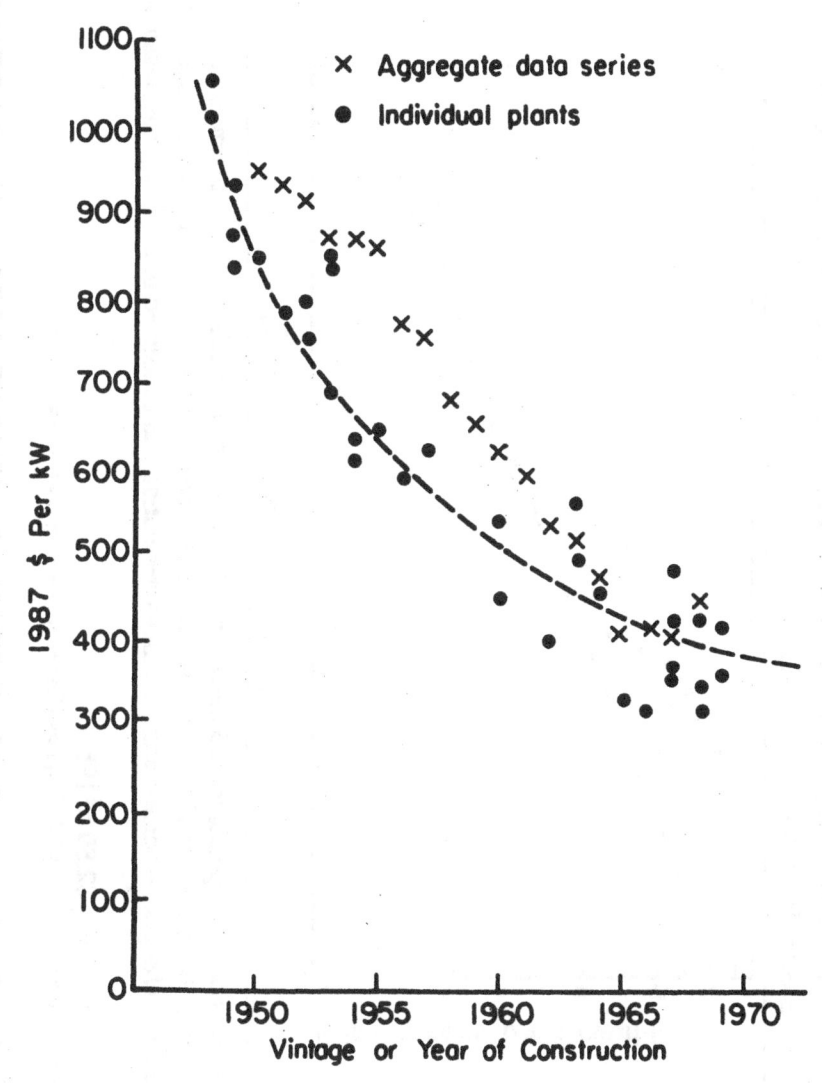

Figure 5.1. Cost of Capacity Additions of Fossil-Fueled Steam-Electric Plants in the US (1987 Dollars per kW)

Source: Hass, J. E., Mitchell, E. J., Stone, B. K. 1974. *Financing the Energy Industry*. Cambridge, Mass.: Ballinger.

Figure 5.2. Average Price of Electricity (1970 Cents/kWh) vs. Cumulative Production of Electricity in the US, 1926 -1970[*]

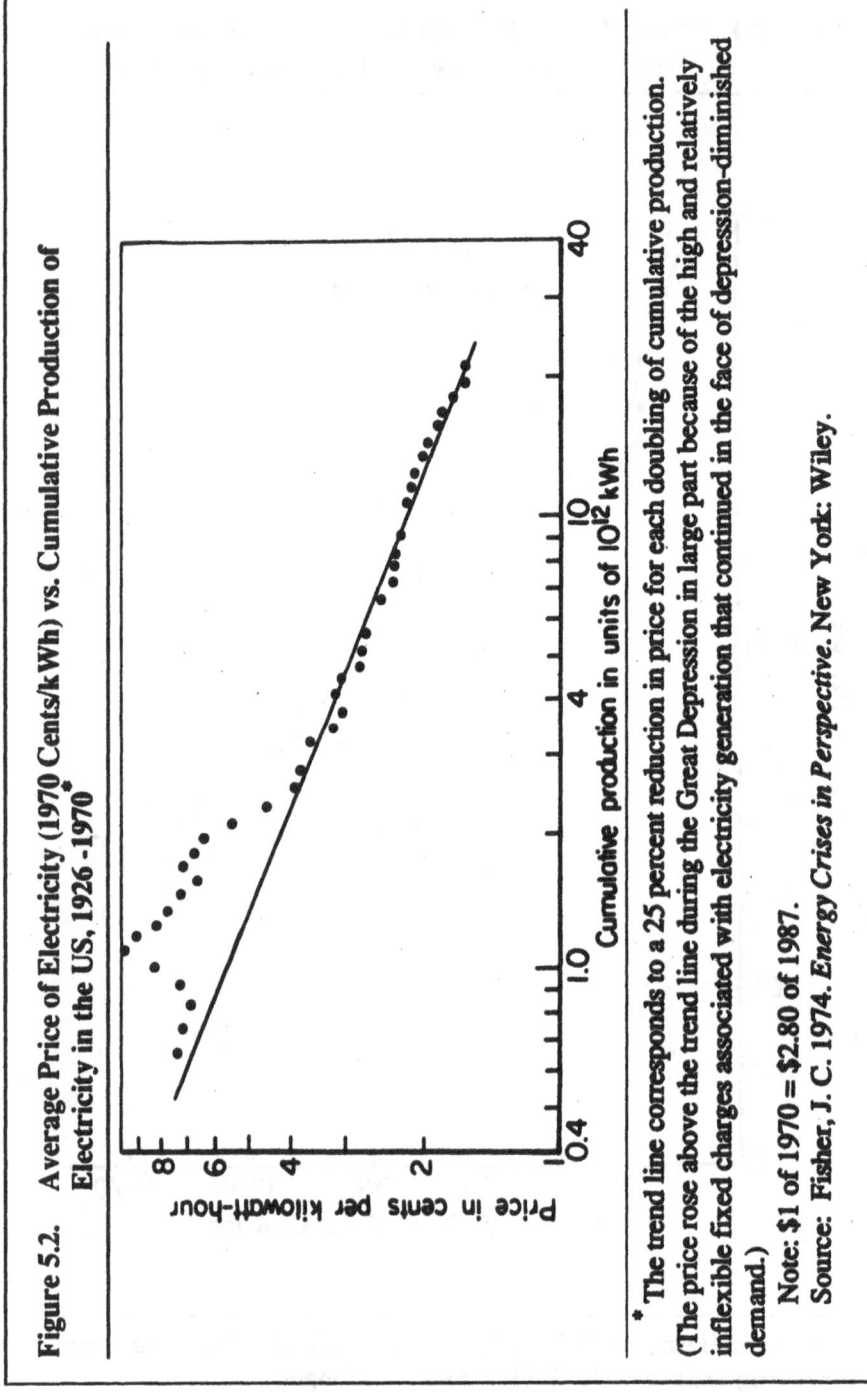

[*] The trend line corresponds to a 25 percent reduction in price for each doubling of cumulative production. (The price rose above the trend line during the Great Depression in large part because of the high and relatively inflexible fixed charges associated with electricity generation that continued in the face of depression-diminished demand.)

Note: $1 of 1970 = $2.80 of 1987.

Source: Fisher, J. C. 1974. *Energy Crises in Perspective*. New York: Wiley.

the average price of electricity in the US today would be 1/3 lower than in 1970. Instead, the average electricity price in the US today is 40% higher than in 1970 or more than double what it would have been if the long-term, cost-cutting trend continued (Figure 5.3).

Escalating capital costs for new central station power plants have contributed greatly to these rising electricity prices. A nuclear plant ordered in the US in 1970 cost about $900 per kW.[1] [Unless explicitly indicated otherwise, all costs and prices are presented here in 1987 dollars.[2]] In a 1985 study by the International Energy Agency (IEA), it is estimated that the cost of a new nuclear plant that would be started up in 1990 in Europe would cost $1900 per kW if construction took ten years or $1700 per kW if construction could be completed in six years.[3] The Electric Power Research Institute (EPRI) has estimated that the cost of a new nuclear plant ordered today in the US would be $3060 per kW, but that this cost could be reduced to $1670 per kW in a "reborn" nuclear industry, i.e., one featuring a streamlined nuclear licensing process, a shorter construction period (six instead of eight years), and improved labor productivity.[4] The costs of coal-fired steam-electric plants with flue gas desulfurization (FGD), which cost about $500 per kW in the US in 1970,[5] have also escalated markedly. The IEA has estimated that the installed cost of a plant with two 600 MW units would be about $1230 per kW in Europe;[6] EPRI has estimated a slightly higher cost ($1340 per kW) for a plant with two 500 MW units in the US.[7]

In a 1987 World Energy Conference (WEC) study,[8] it is projected that the ongoing escalation in power generation costs will continue (Table 5.1). For industrial countries, it is projected that the share of GDP that will have to be spent on electricity supply expansion will be in the range 2.0% to 2.7% of GDP in 2000, compared to 2.2% in 1980, even though electricity demand is projected to grow only 0.65 to 0.83 times as fast as GDP, 1980-2000. For developing countries, where electricity demand growth is expected to be more rapid, it is projected that the share of GDP that will have to be spent on electricity supply expansion will increase from 1.5% of GDP in 1980 to the range of 2.6% to 5.5% in 2000. For the already capital-constrained developing countries, this means that supporting electricity demand growth rates that are modest by historical standards will be extremely difficult.[9]

The Prospects for Improving Steam-Electric Power Technology

Before discussing the opportunities for innovation afforded by advanced gas turbine technologies, let us examine the prospects for

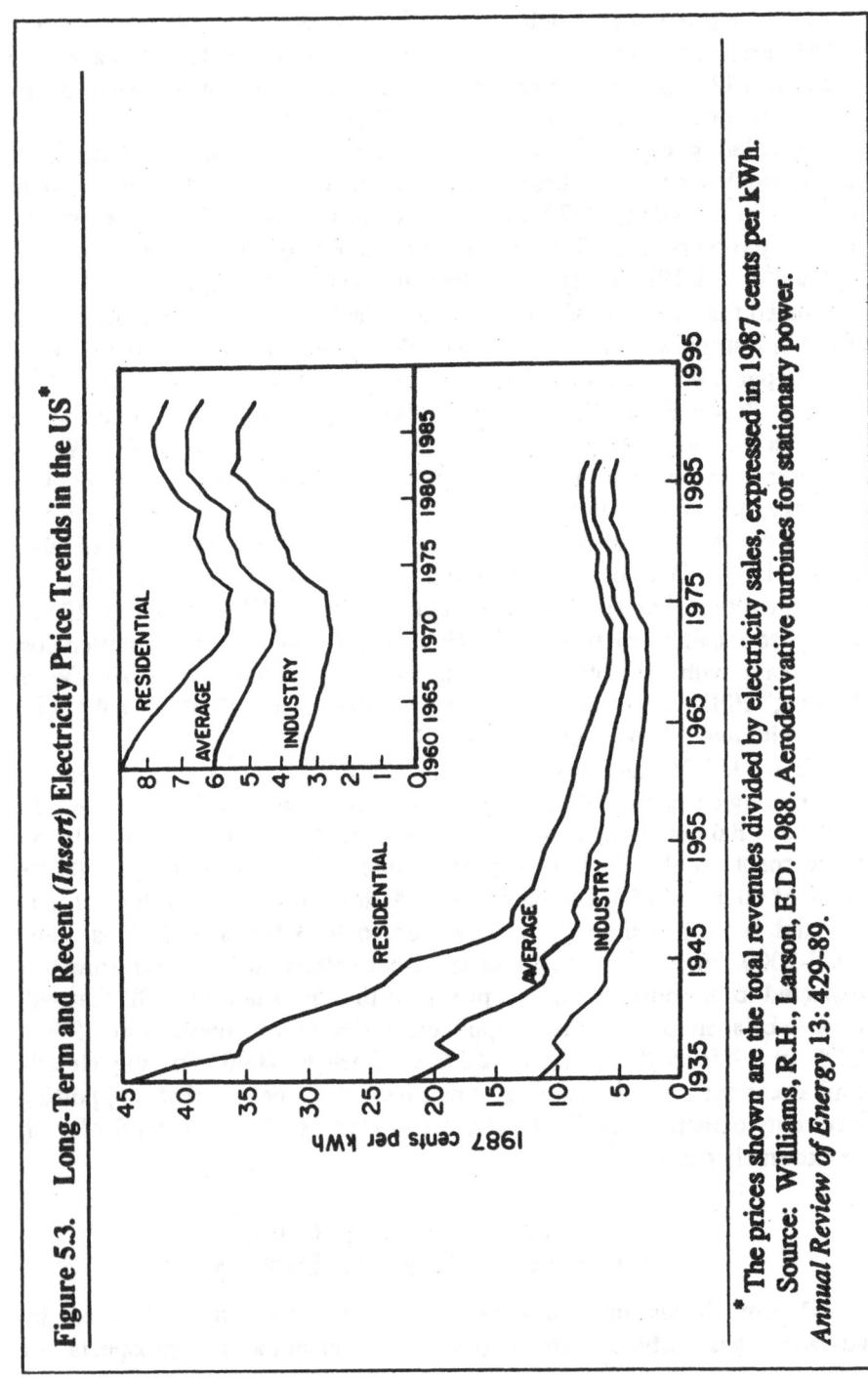

Figure 5.3. Long-Term and Recent (*Insert*) Electricity Price Trends in the US[*]

[*] The prices shown are the total revenues divided by electricity sales, expressed in 1987 cents per kWh.
Source: Williams, R.H., Larson, E.D. 1988. Aeroderivative turbines for stationary power.
Annual Review of Energy 13: 429-89.

Table 5.1. Capital for Electricity[a]

	1980	2000L[b]	2000H[b]
Cost for New Generating Capacity ($/kW)			
Hydroelectric	2740	3360	4110
Nuclear	2060	2540	3080
Fossil Fuel, Thermal	1030	1230	1510
Average Capital Requirements by Region ($/kW)			
Industrial Countries			
Generation	1480	2000	2390
T & D	2740	2770	3030
Developing Countries			
Generation	1690	2070	2480
T & D	810	1700	2480
Centrally Planned Economies			
Generation	1370	1810	2320
T & D	1370	1960	2620
Overall Capital Requirements for Electricity [10^9 $/year (% of GDP)]			
Industrial Countries	226 (2.2)	302 (2.0)	488 (2.7)
Developing Countries	44 (1.5)	148 (2.6)	381 (5.5)
Centrally Planned Economies	60	147	233

	1980-2000L[b]	1980-2000H[b]
Average Growth Rates (%/Year)		
For GDP		
Industrial Countries	2.0	3.0
Developing Countries	3.5	4.5
For Primary Energy Consumption		
Industrial Countries	0.15	1.3
Developing Countries	2.5	4.7
Centrally Planned Economies	1.8	2.3
For Electricity Generation		
Industrial Countries	1.3	2.5
Developing Countries	4.5	6.8
Centrally Planned Economies	2.7	3.2

[a] According to a 1987 World Energy Conference (WEC) study (Schneider, H.K. 1987. *Investment Requirements of the World Energy Industries 1980-2000*. London: World Energy Conference).

[b] 2000L (2000H) is for the WEC low (high) growth scenario.

improving the technology for nuclear and fossil fuel-based steam-electric power generation.

The cost escalations plaguing steam-electric power generation are due in part to tightening environmental and safety rules. Other important factors include inadequate quality control in equipment manufacture and construction, bottlenecks that have arisen because each big project has been in many ways unusual, and escalating labor costs arising from shortages of qualified manpower and declines in labor productivity. Many such problems result in not only direct cost increases, but also indirect cost increases associated with the accumulated interest charges from extended construction periods.

A 1974 analysis by John Fisher of the escalation in nuclear power costs in the decade leading up to the first oil crisis provides an important insight relating to these construction-related problems that seems relevant for power cost escalation generally since 1970:[10]

"When measured in constant dollars per kilowatt of capacity, the cost of constructing a nuclear power plant increased by perhaps 50 percent in the past decade... When power plant costs rise an explanation is required, as we expect all power plant costs to decline through the economies of scale and new technology. The environmental movement was responsible for part of the rise in nuclear plant costs, by causing various procedural delays and by requiring additional expensive safeguards to protect against hypothetical accidents. But there appears to be another cause for increasing construction costs, associated with a growing portion of high-cost field construction and a shrinking proportion of low-cost factory construction for the very large power plants now being built... the costs associated with a shift to field from factory can more than offset anticipated economies of scale..."

Fisher pointed out that historically, as electric utility plant capacity doubled every decade, factory capacity also doubled, as did field construction at each site. Manufacturing and construction costs per kW declined in the factory and in the field, since each of these increased its scale of operations. As long as both activities grew in proportion, the economies of scale produced similar cost reductions in each, and therefore an overall cost reduction, even though the unit cost of field construction was always higher than the unit cost of factory construction. This pattern held until plant size reached about 200 MW. Then, because design engineers felt that scale economies would be much more important for nuclear than for fossil fuel plants, nuclear power plant capacities were built in sizes of the order of 1000 MW -- shifting a greater portion of the construction from the factory to the field, upsetting the pattern of the past, with the result that a much larger fraction of the construction was carried out at smaller, less-efficient field locations. Fisher's important insight is that the widely touted economies of scale in power plant construction are illusory because: (a) field construction is inherently more costly than factory construction, and (b) with field construction it is never possible to get very far down the "learning curve," in contrast to the situation with factory production. This "diseconomy of scale" problem has persisted to the present.

The increase in scale for coal-fired power plants to the present range of 500 to 600 MW was in part a competitive response to large-scale nuclear plant construction. Also, the pursuit of scale economies was seen as an opportunity for continuing the historical reductions in the cost of electricity (Figures 5.2 and 5.3), after the long term trend toward increased thermodynamic efficiency of power generation ceased in the late 1950s (Figure 5.4). But as in the case of nuclear power, the point of diminishing returns to scale appears to have been reached or exceeded for coal plants. Even where scale economy gains might be realized in construction, these gains tend to be offset by losses in reliability for the larger units.[11]

What are the prospects for improving thermodynamic efficiency? Since the 1920s, most gains in efficiency in steam-electric power plants have been due largely to increases in maximum steam temperatures and pressures. By the 1950s, peak temperatures had reached 565°C and peak pressures 165 bar for subcritical steam units and 241 bar for supercritical steam units.

There are ongoing developmental efforts aimed at improving steam-electric power plant efficiency by increasing peak steam temperatures.[12,13] But as peak steam temperatures are increased, problems of materials strength, oxidation, and corrosion rapidly become more serious, dictating shifts to more costly high-strength, oxidation and corrosion-resistant alloys for the large steam-tubing heat exchangers that transfer heat from the combustor to steam at high temperature and pressure. (See, for example, Figure 5.5, which shows, for a number of alloys used in steam tubing exposed to high temperatures, that the maximum allowable stress declines rapidly beyond a critical threshold.)

Peak steam temperatures have not increased since the 1950s, and in fact utilities today tend to choose a slightly lower peak temperature of 540°C in coal plants. They do so not only because of the lower capital cost, but also because, even with judicious choice of better tubing materials, higher temperature operating conditions have led to more forced outages, owing to tubing damage from problems such as coal-ash corrosion.[14]

A 1976 Westinghouse study, the results of which are consistent with many other studies carried out since the 1950s,[15] indicates the magnitude

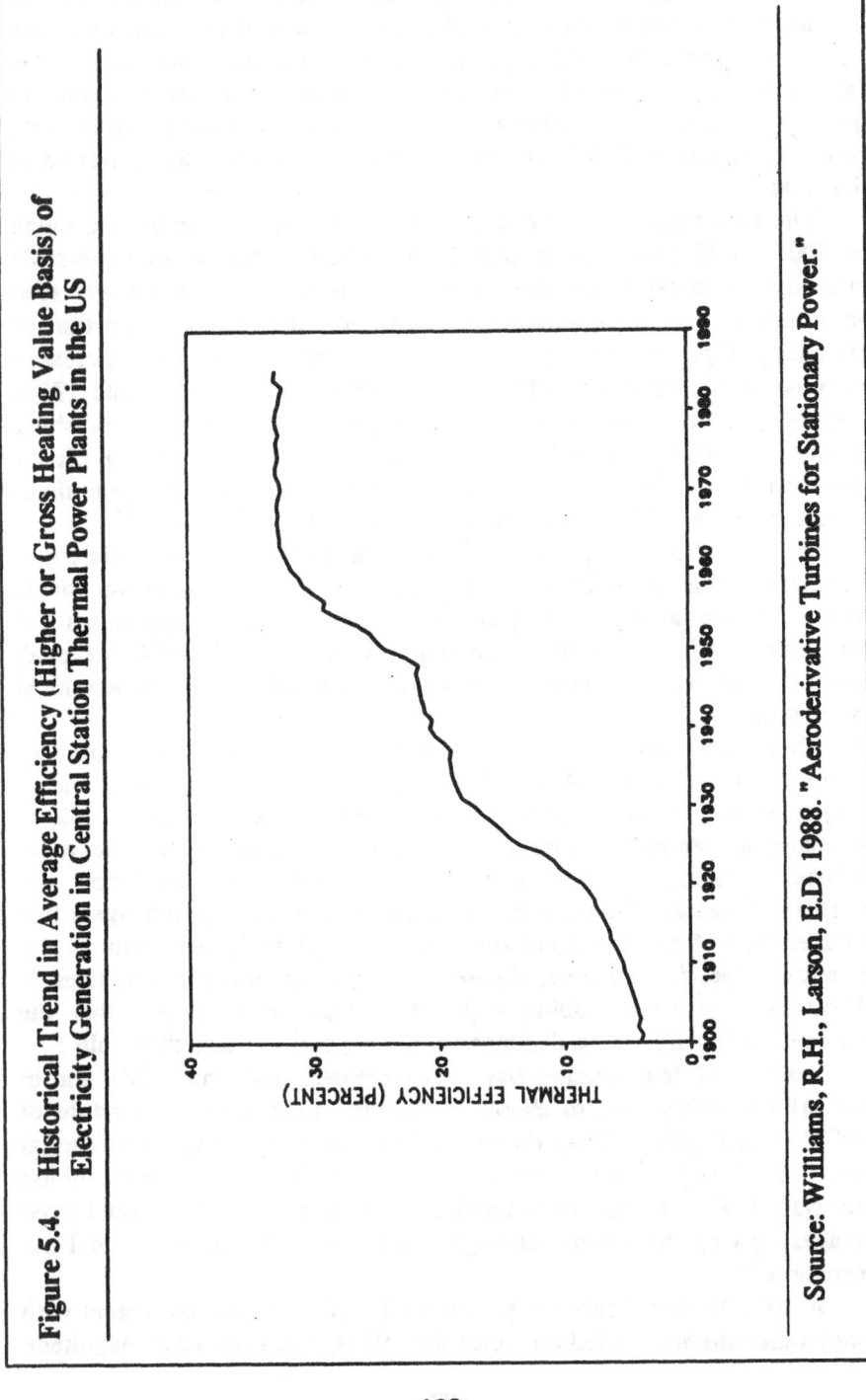

Figure 5.4. Historical Trend in Average Efficiency (Higher or Gross Heating Value Basis) of Electricity Generation in Central Station Thermal Power Plants in the US

Source: Williams, R.H., Larson, E.D. 1988. "Aeroderivative Turbines for Stationary Power."

Figure 5.5. **Effect of Temperature on Maximum Allowable Stress for Different Steel Alloys Used for Steam Tubing in High Temperature Service**[*]

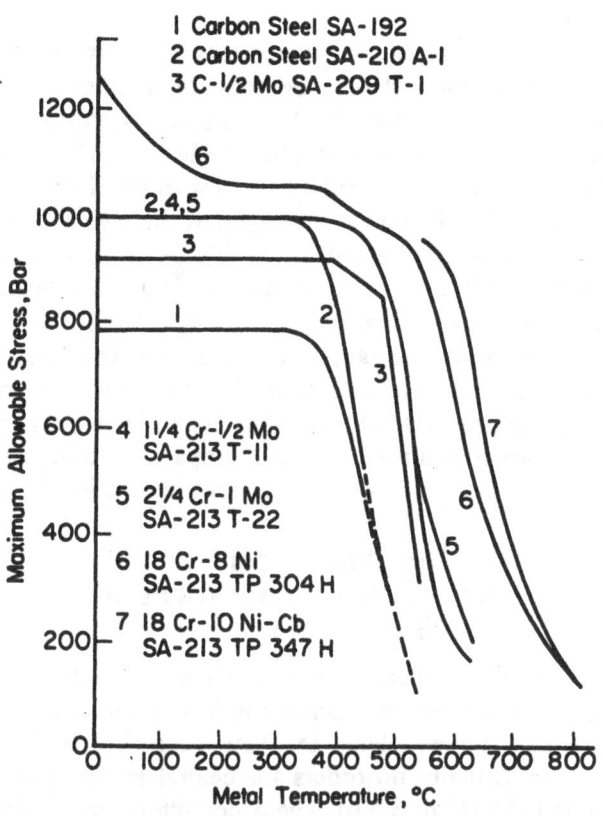

*According to the Boiler Code of the American Society of Mechanical Engineer (ASME).

Notes: 1 = low strength carbon steel; 2 = intermediate strength carbon steel; 3 = a ferritic alloy containing 0.5 percent molybdenum; 4 = ferritic alloy containing 1.25 percent chromium and 0.5 percent molybdenum; 5 = ferritic alloy containing 2.25 percent chromium and 1.0 percent molybdenum; 6 = austenitic stainless alloy containing 18 percent chromium and 8 percent nickel; 7 = austenitic stainless alloy containing 18 percent chromium and 10 percent nickel.

Source: Combustion Eng., Inc. 1981. *Combustion: Fossil Power Systems.* Windsor, Conn.: Rand McNally.

of the tradeoff involved in increasing the maximum steam temperature of a 500 MW steam plant; an increase from 540 to 650°C would increase the plant efficiency by 6%, but at the cost of a 26% increase in capital cost.[16] On a lifecycle cost basis, the price of coal would have to increase from $50 to $200 per tonne before it would be worthwhile to shift to the higher peak steam temperature.

While the outlook for major improvements in steam-electric power technology is not auspicious, it may be feasible to increase efficiency without pushing peak working fluid temperatures further, through development of the recently proposed Kalina cycle.[17,18] The Kalina cycle is a novel modified Rankine cycle that uses as a working fluid a mixture of ammonia and water that is varied throughout the cycle. A 3 MW demonstration plant is being planned at the US Department of Energy's Engineering Center in Canoga Park, California.[19] The big uncertainties regarding the Kalina cycle are the complicated "plumbing" and possible difficulties associated with managing the binary working fluid at high temperatures and pressures, which might lead to capital and operating and maintenance cost penalties. Also, the performance estimates for the Kalina cycle are for very small assumed pressure drops and tight temperature differences, conditions that are difficult to achieve in practice.[20]

The Outlook for
Stationary Power Applications of
Gas Turbines

Rising costs of steam-electric power plants and slow, uncertain electrical load growth have led to dramatic reductions in the construction of new central station power plants in many countries; in the US, for example, there were virtually no orders for central station power plants between 1982 and 1986 (Figure 5.6). These conditions are leading utility planners to give more attention to the gas turbine for meeting future power needs. On the basis of contacts made with the majority of the large electric utilities, Gluckman at the Electric Power Research Institute estimated that some 40 GW of new gas turbine-based generating capacity is on order or being planned in the US for installation by 1995.[21]

Emerging utility interest in gas turbines is complemented by a boom in gas turbine sales to cogenerators in the US. Electric utility rate increases and the Public Utility Regulatory Policies Act of 1978 (PURPA), along with the 1982 and 1983 Supreme Court decisions upholding its provisions, have led to a competitive challenge for utilities from independent cogenerators and small power producers in the US PURPA encourages cogeneration and the production of electricity from renewable

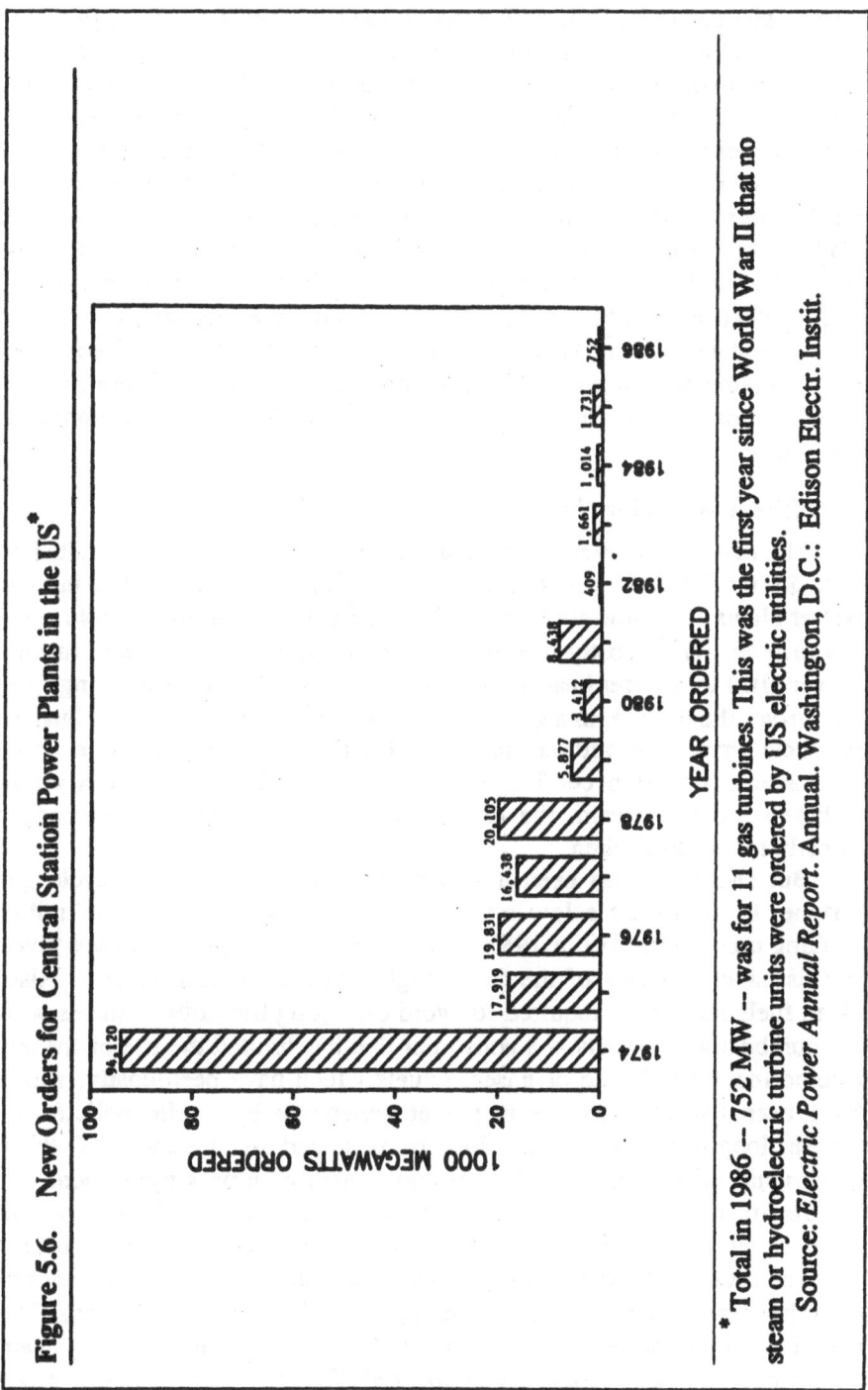

Figure 5.6. New Orders for Central Station Power Plants in the US*

* Total in 1986 -- 752 MW -- was for 11 gas turbines. This was the first year since World War II that no steam or hydroelectric turbine units were ordered by US electric utilities.

Source: *Electric Power Annual Report.* Annual. Washington, D.C.: Edison Electr. Instit.

energy sources in small installations by requiring utilities: (a) to purchase the electricity from qualifying producers at a price equal to the cost the utility can avoid by not having to otherwise supply that power and (b) to provide back-up power at reasonable rates. Between 1980 and 1987, 62 GW of electrical generating capacity were certified as qualifying for PURPA benefits by the Federal Energy Regulatory Commission (FERC), nearly three-fourths of which is due to cogeneration (Figure 5.7). Some 20 GW of the cogeneration capacity certified in this period, more than the sum of all utility orders for all kinds of central station power plants, 1980-1986 (Figure 5.6), was based on the gas turbine (Figure 5.7).

Recent interest in the gas turbine for stationary power reflects both long-standing attractions of this technology and recent improvements that make it possible for the gas turbine to compete in a wider range of markets.

Traditional Roles for Gas Turbines

The historical attraction of the gas turbine for utilities has been its low cost, \$300 per kW[22] or less, a small fraction of the cost of coal or nuclear power plants. This low cost reflects the utter simplicity of the simple cycle gas turbine. While costly heat exchangers are required in a steam turbine power plant to transfer heat from the combustor to the steam working fluid that drives the turbine, in a gas turbine power plant the hot fuel combustion products drive the turbine directly (Figure 5.8a). Also, while large condensers and often cooling towers are required to condense a steam turbine's exhaust steam, the exhaust from a gas turbine is discharged directly to the atmosphere.

But simplicity has been a mixed blessing for the simple cycle gas turbine. It has meant a low efficiency; the average efficiency of utility peaking units in the US in 1985 was only 29%.[23] [Efficiencies are given in this paper in terms of the fuel's higher (gross) heating value.] Also, clean fuels have been required to avoid damaging the turbine blades with the combustion products -- a constraint that has limited the use of the gas turbine mainly to liquid or gaseous fuels which have been costly, which have been barred from use in power generation by public policies, or whose long-term availability has been uncertain. Because of these constraints, utilities have used gas turbines mainly for peaking service.

Its low unit capital cost has also helped make the gas turbine attractive for cogeneration applications. Because of the relative insensitivity of gas turbine unit costs to scale (Figure 5.9), the gas turbine tends to be favored over the steam turbine for all but the largest cogeneration installations. The use of the high-temperature (425 to 540°C) turbine exhaust to raise steam in a heat recovery steam generator (HRSG) for heating applications

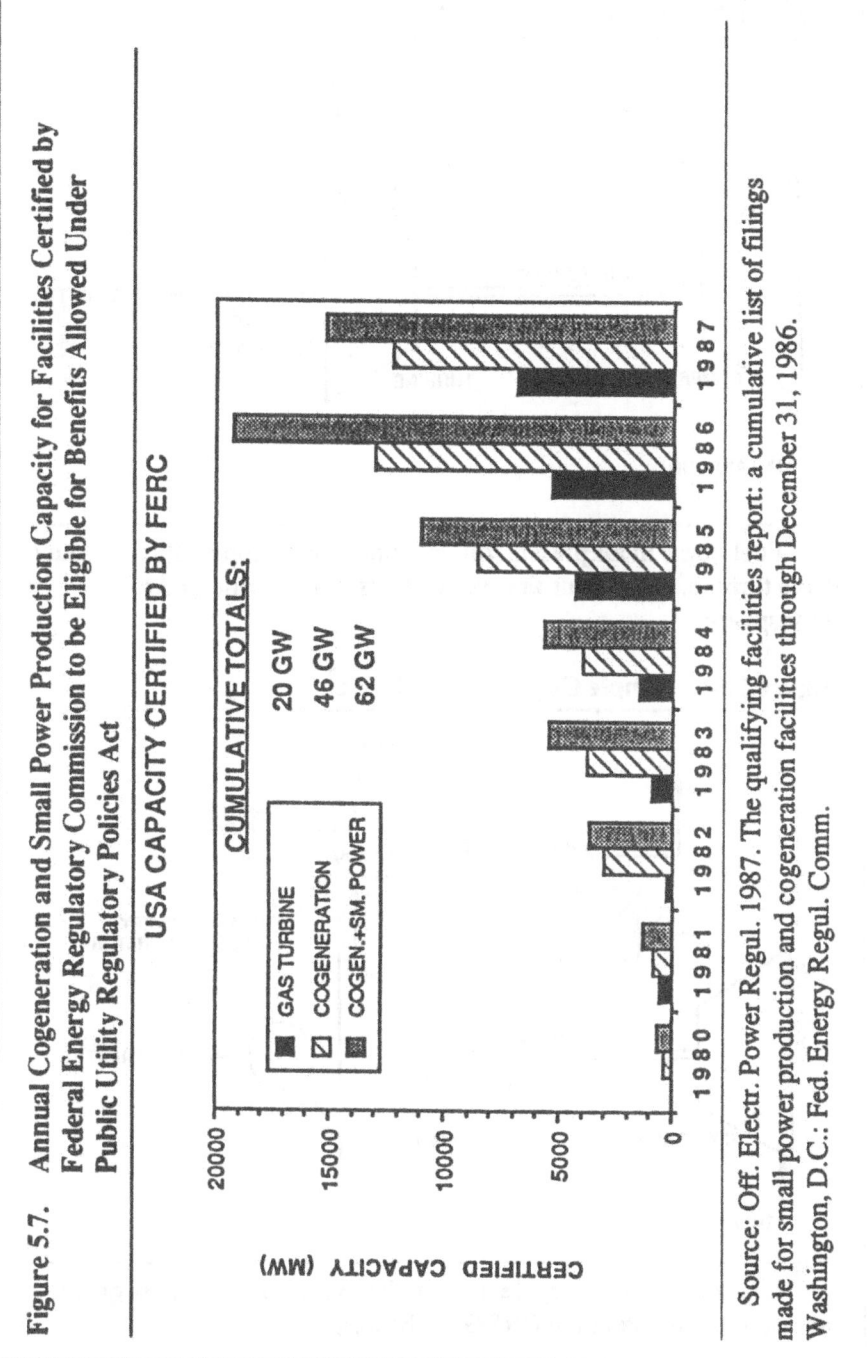

Figure 5.7. Annual Cogeneration and Small Power Production Capacity for Facilities Certified by Federal Energy Regulatory Commission to be Eligible for Benefits Allowed Under Public Utility Regulatory Policies Act

USA CAPACITY CERTIFIED BY FERC

CUMULATIVE TOTALS:

GAS TURBINE   20 GW
COGENERATION   46 GW
COGEN.+SM. POWER   62 GW

Source: Off. Electr. Power Regul. 1987. The qualifying facilities report: a cumulative list of filings made for small power production and cogeneration facilities through December 31, 1986. Washington, D.C.: Fed. Energy Regul. Comm.

Figure 5.8a. Simple Power Cycle[*]

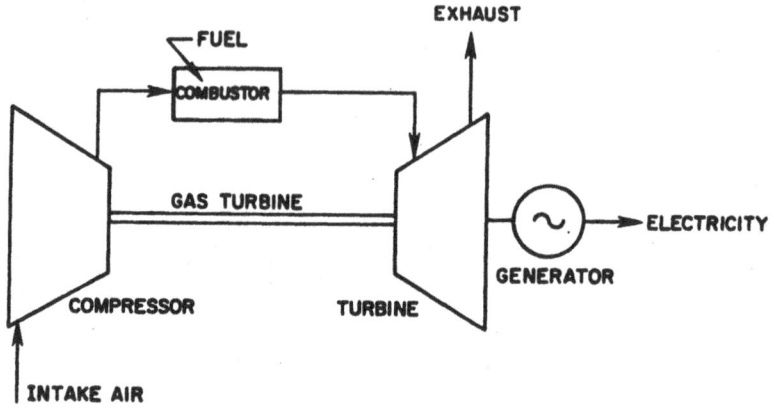

[*] Fuel burns in air pressurized by compressor, combustion products drive turbine, and hot turbine exhaust gases are discharged to atmosphere.

Figure 5.8b. Simple Cogeneration Cycle[*]

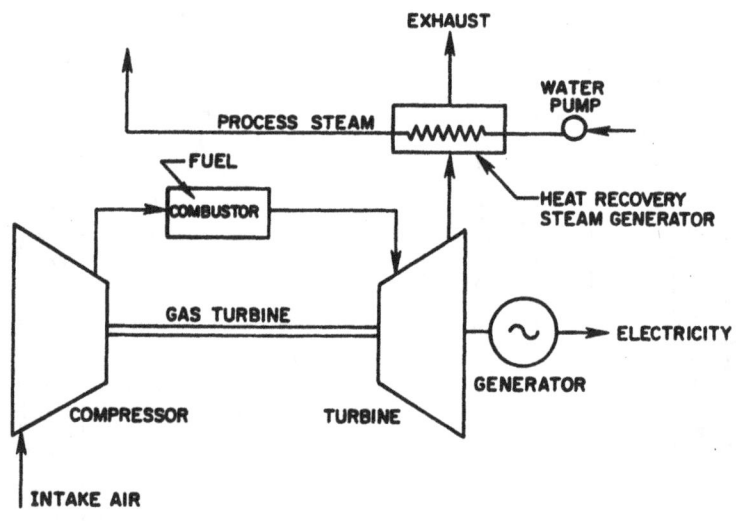

[*] Like simple power cycle, except that hot turbine exhaust gases are used to raise steam in HRSG for heating.

Figure 5.8c. Combined Cycle for Power*

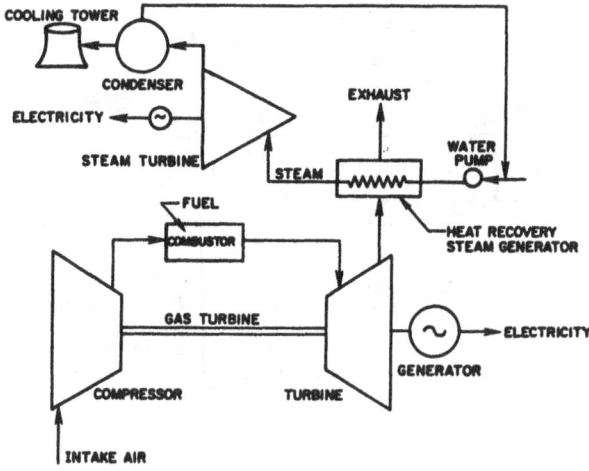

---

*Like simple cogeneration cycle, except that steam from HRSG is used to produce extra power in condensing steam turbine.

Figure 5.8d. Combined Cycle for Cogeneration*

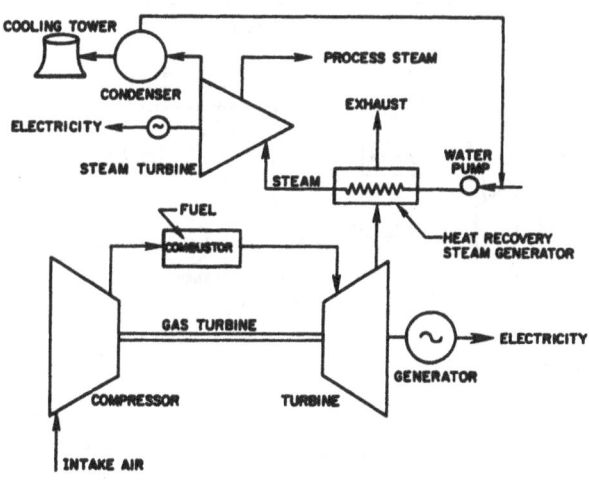

---

*Like combined cycle for power, except that some steam is bled from steam turbine for heating.

Figure 5.8e.   Steam-Injected Gas Turbine (STIG)[*]

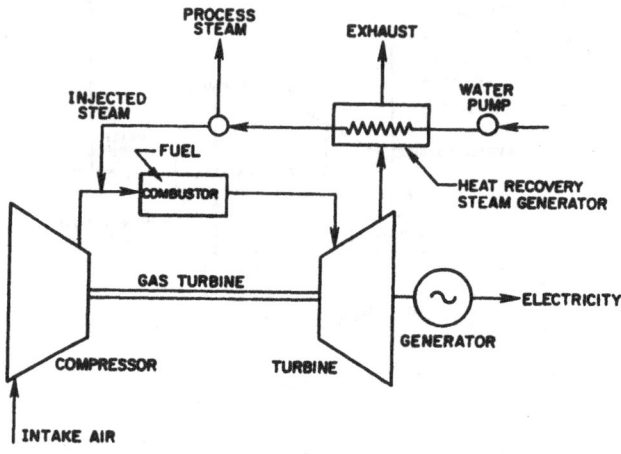

[*] Like simple cogeneration cycle, except that steam not needed for heating is injected into combustor for increased power output and higher electrical efficiency.

Figure 5.8f.   Intercooled Steam-Injected Gas Turbine (ISTIG)[*]

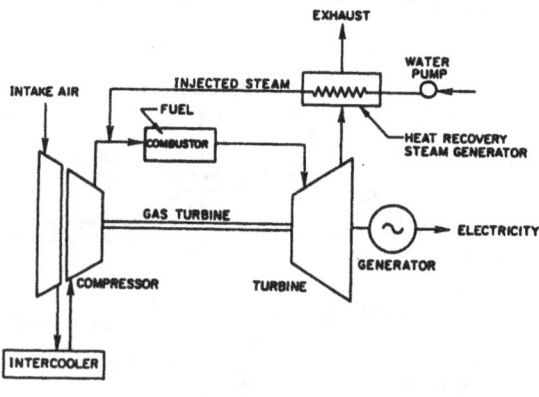

[*] Like STIG with full steam injection except that intercooler between compressor stages allows for operation at much higher turbine inlet temperature because of improved air cooling of turbine blades.

Figure 5.9. Unit Installed Costs for Small-Scale Cogeneration Systems *

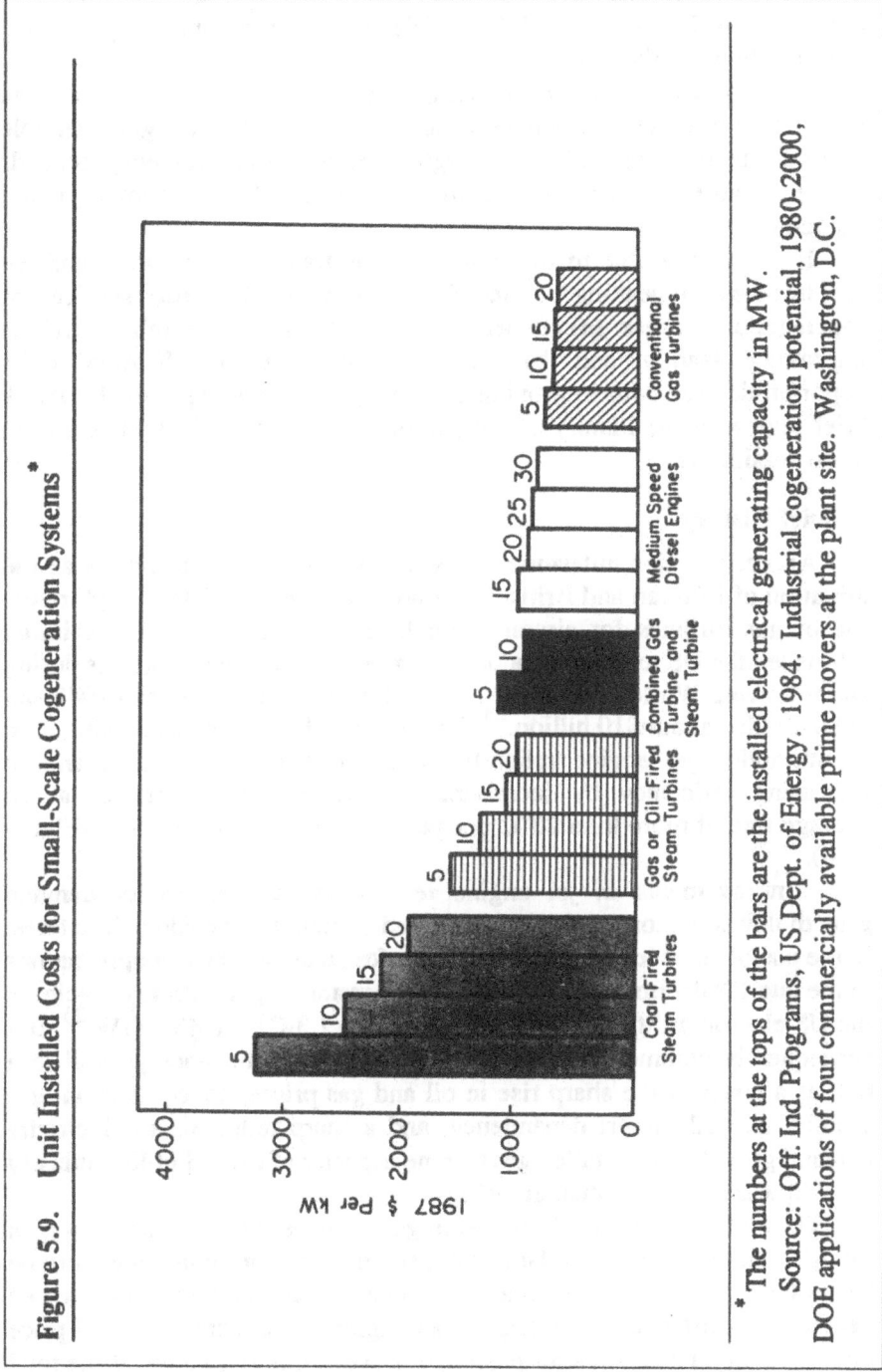

* The numbers at the tops of the bars are the installed electrical generating capacity in MW.

Source: Off. Ind. Programs, US Dept. of Energy. 1984. Industrial cogeneration potential, 1980-2000, DOE applications of four commercially available prime movers at the plant site. Washington, D.C.

(Figure 5.8b) makes the gas turbine a thermodynamically efficient cogeneration device, even if the efficiency of the turbine for producing power only is relatively poor.

A major shortcoming of the simple cycle gas turbine in cogeneration is that it is often uneconomical in applications involving highly variable steam loads, because achieving a high thermodynamic efficiency depends on being able to make use of the full electrical and thermal output capacities.

It is now possible to overcome the constraints which have confined gas turbines to peaking service for utilities and baseload service for cogeneration because: (a) the performance of the basic gas turbine cycle is improving steadily, and (b) various simple cycle modifications offer opportunities for both improving efficiency and reducing capital cost. A brief review of the history of the gas turbine is helpful in understanding these possibilities.

A Brief History

An early major milestone in the history of the gas turbine was the initiation of German and British programs in the mid-1930s to explore the use of gas turbines for aircraft propulsion. The success of these initial efforts led the US to launch major jet engine development programs during and following World War II: the cost of these programs between 1940 and 1980 totalled about $10 billion.[24] These efforts have been successful, both in improving jet engine reliability and thrust-to-weight ratios and in increasing efficiency by increasing turbine inlet temperatures, at an average rate of more than $20^{\circ}C$ per year, between 1950 and 1980 (Figure 5.10).

Improvements in jet engine technology and electricity demand growth that was more rapid than expected stimulated considerable interest in the use of short lead-time gas turbines for stationary power applications in the late 1960s. Between 1965 and 1975, installed gas turbine capacity in the US electric utility industry increased from 1.3 GW to 43.5 GW.[25] But subsequently commercial interest in stationary gas turbines ground to a halt as a result of the sharp rise in oil and gas prices, concerns about gas scarcity and oil import dependency, and a sharp reduction in electricity demand growth; the installed gas turbine capacity of the US utility industry in 1985 was no greater than in 1975.[26]

The end of commercial interest in gas turbines for stationary power in the US did not slow fundamental progress in improving gas turbine technology, however. One reason is that commercial airlines pressed vendors to improve the efficiency of jet engines. The rising world oil price increased the fuel costs of air passenger travel from 11 to 32% of the total

138

Figure 5.10. Trend in Turbine Inlet Temperatures for Advanced Aircraft Jet Engines and Long-Life Industrial Turbines *(Left)* and Turbine Blade Material Operating Temperature *(Right)*.

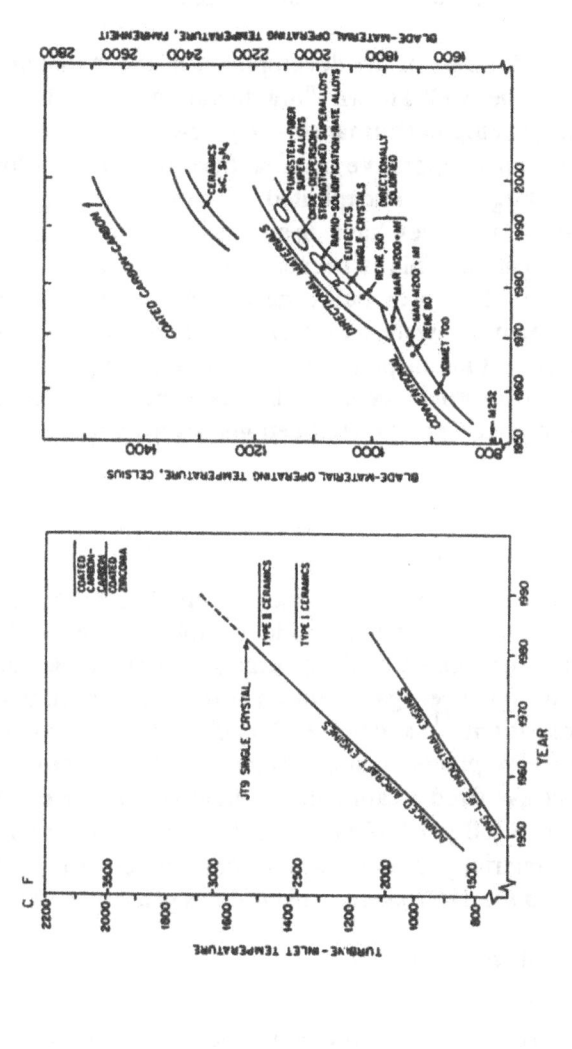

Note: When an aircraft engine is modified for stationary applications the rated turbine inlet temperature is reduced about 110°C to promote long-life operation.

Source: Wilson, D.G. 1984. *The Design of High-Efficiency Turbomachinery and Gas Turbines.*

cost of air passenger service in the 1970s.[27] US Department of Defense support for research and development on jet engines for military applications also continued at a high level, averaging about $450 million per year in the decade ending in 1986.[28] Continuing R&D in this area is expected to bring significant further increases in turbine inlet temperatures by the turn of the century, as a result of major improvements in blade materials (Figure 5.10) and more effective blade-metal cooling technologies.[29]

A paradoxical aspect of the development of the stationary gas turbine is that most of the well-known "low-technology" cycle modifications available for improving performance -- e.g., steam injection, intercooling, reheat, regeneration, evaporative regeneration, and steam-reforming of the fuel--remain largely unexploited, even though enormous "high-technology" advances have been made in turbine blade materials, design, and fabrication. This is because such cycle modifications involve the introduction of heavy or bulky heat exchangers or the use of large quantities of steam or water, neither of which is relevant to aircraft applications. This situation presents an enormous opportunity because it means that major improvements can be made in the performance of gas turbines for stationary power applications with relatively modest R&D efforts.

Progress in
Combined Cycle Technology

One gas turbine cycle modification is familiar to the electric utility industry: the gas turbine/steam turbine combined cycle (Figure 5.8c), which accounted for some 4.7 GW of utility generating capacity in the US in 1985.[30] With advanced gas turbines now commercially available for stationary applications,[31] a combined cycle efficiency of 45% can be realized in 200 MW plants costing $520 per kW. For comparison, a 36% efficient natural gas-fired steam-electric plant consisting of two 500 MW units would cost $760 per kW (Table 5.2). These combined cycle plants could produce electricity at a busbar cost only three-fifths of that for large new coal plants at the 1987 average natural gas price in the US (Table 5.2).

A Comparison of Industrial and
Aeroderivative Turbines

The combined cycle is a good technology for beginning a transition to greater use of gas turbines in stationary power applications. It marries the new gas turbine technology to the familiar steam turbine. But it would be a mistake to limit utility use of advanced gas turbines to this option, because alternatives may offer advantages in some applications.

In exploring alternative gas turbine strategies, it is useful to distinguish between the characteristic features of heavy-duty industrial turbines and aeroderivative units. Various vendors offer heavy-duty industrial units in sizes ranging up to 70 to 135 MW. The tendency has been to design them with modest compression ratios (8 to 16). They are thus well-suited for combined cycle operations because the turbine exhaust gases are thereby relatively hot -- 593°C for the most recently offered advanced industrial unit[32] -- making it possible to produce high quality steam in the heat recovery steam generator (HRSG).

In contrast, aeroderivative units are lightweight and compact, with relatively small capacities -- 30 to 35 MW at the high end of available capacities -- and the trend is toward high compression ratios (18 to 30); all such characteristics reflect jet engine design requirements. Though simple cycle aeroderivatives are relatively efficient as electricity producers, such engines tend to be poor candidates for combined cycle applications, since the turbine exhaust gases are not nearly as hot as in heavy-duty industrial units. Until recently this attribute led to the neglect of aeroderivatives as a serious candidate for central station power generation. However, recent developments have shown that aeroderivatives are good candidates for other efficiency and output-augmenting cycle modifications such as steam injection, discussed below.

While aeroderivative turbines are not nearly as familiar as heavy-duty industrial turbines for stationary applications, they warrant attention not only because, with appropriate cycle modifications, they can perform as well as or better than industrial units, but also because aeroderivatives have some other important attributes.

First, aeroderivatives can bring back to power generation the advantages of cost-cutting mass production. Moreover, when an aeroderivative is introduced for power generation, it is already well-advanced on the learning curve, because large-volume production of jet engines for aircraft applications has preceded it.

Second, the aeroderivatives are expected to benefit from continuing advances in jet engine technology, which can be transfered quickly and at low incremental cost to stationary applications. While there are only modest ongoing development efforts to improve industrial turbines in the US, there is continuing heavy US government support for jet engine R&D. This includes, for example, the new $3.4 billion, 13-year Integrated High Performance Engine Technology program supported by the Department of Defense and the National Aeronautics and Space Administration.[33] Such R&D efforts are expected to lead to major improvements in aircraft engine technology, including substantial further increases in turbine inlet temperatures (Figure 5.10).

Table 5.2. Cost/Performance Characteristics for U.S. Central-Station Power Plants[a]

	Coal and Nuclear Steam-Electric Plants				
	Coal[b,c]			Light Water Reactor[d]	
				Current	Target
Unit Size (MW)	2 x 500	500	200	1100	1100
Efficiency (%)[e]	34.6	34.6	34.6	33.4	33.4
Unit Cost ($/kW)	1340	1410	1880	3060	1670
Levelized Busbar Cost (Cents/kWh)					
Capital[f]	1.61	1.69	2.25	3.66	2.00
Fuel	1.86	1.86	1.86	0.91	0.91
O&M	0.89	0.99	1.37	1.11	1.11
Total	4.36	4.54	5.48	5.68	4.02

	Natural Gas-Fired Plants[g]							
	1987 Natural Gas Price[h]				2 X 1987 Natural Gas Price			
	Steam	ACC	STIG	ISTIG	Steam	ACC	STIG	ISTIG
TIT (°C)	540	1260	1200	1370	540	1260	1200	1370
Unit Size (MW)	2 x 500	205	4 x 51	114	2 x 500	205	4 x 51	114
Efficiency (%)[e]	36.3	45.0	40.0	47.03	6.3	45.0	40.0	47.0
Unit Cost ($/kW)	760	520	410	400	760	520	410	400
Levelized Busbar Cost (Cents/kWh)								
Capital[f]	0.91	0.63	0.49	0.48	0.91	0.63	0.49	0.48
Fuel	2.08	1.68	1.89	1.61	4.17	3.36	3.78	3.22
O&M	0.49	0.29	0.29	0.29	0.49	0.29	0.29	0.29
Total	3.48	2.60	2.67	2.38	5.57	4.28	4.56	3.99

[a] All costs are in 1987 U.S. dollars.

[b] Capital costs, efficiencies, and O&M costs are EPRI estimates, for a bituminous coal-fired subcritical steam plant with flue gas desulfurization, according to EPRI, TAG 1986 (Electr. Power Res. Inst. 1986. Technical Assessment Guide. 1: Electricity Supply -- 1986. Palo Alto, Calif.).

[c] The assumed coal price is $1.79/GJ, the average U.S. utility price projected for 2000 by the U.S. Department of Energy [Energy Inf. Admin. 1988. *Annual Energy Outlook 1987*, with projections to 2000. DOE/EIA-0383(87). Washington, D.C.: U.S. GPO].

(Table 5.2 notes cont'd. next page)

Table 5.2 Notes (Cont'd.)

[d] Reactor plant size, unit capital costs, and efficiencies are EPRI estimates (EPRI, TAG 1986). The two sets of capital costs are the current cost and an EPRI target for "improved" conditions -- resulting from higher construction labor productivity, shorter construction period, streamlined licensing process, etc. The assumed nuclear fuel cycle cost is $0.84/GJ, EPRI's projection for the period 1990-2000 (EPRI, TAG 1986). The assumed O&M cost is the 1985 U.S. average for nuclear power plants [Energy Inf. Admin. 1987. Historical plant cost and annual production expenses for selected electric plants 1985. DOE/EIA-0455(85). Washington, D.C.: U.S. GPO], twice as large as the EPRI estimate for new plants (EPRI, TAG 1986).

[e] Based on the fuel's higher heating value and for operation at 100% load.

[f] For a 6.1% real discount rate [recommended by EPRI (EPRI, TAG 1986)], a 30-year plant life, and a 70% capacity factor. No taxes or tax incentives are included.

[g] The steam plant involves 165 bar, 540°C steam (with single reheat to 540°C) driving a conventional turbine/electric generator (EPRI, TAG 1986). The advanced combined cycle (ACC) is a recently commercialized 135 MW GE Frame 7F gas turbine plus a 70 MW steam turbine; the indicated performance is a General Electric estimate (Brandt, D.E. 1986. Heavy-duty turbopower: the MS7001F. Mech. Eng. July: pp. 28-36). The STIG unit is a commercial steam-injected gas turbine based on the GE LM 5000 (L. Gelfand, Manager, Advanced Programs and Ventures, General Electric Marine and Industrial Division, Cincinati, Ohio, personal communication, February 1987). The ISTIG unit is a proposed intercooled steam-injected gas turbine, based on the LM 8000 [Eng. Dept., Pacific Gas and Electr. Co. 1984. Scoping study: LM5000 steam-injected gas turbine. Based on work performed by the Mar. and Ind. Engine Proj. Dept. of the Gen. Electr. Co.; Homer, M.W. (Mar. and Ind. Eng. and Serv. Div. of the Gen. Electr. Co.). 1988. Position statement - intercooled steam-injected gas turbine. Testimony presented at the Committee Hearing for the 1988 Electricity Report of the Calif. Energy Comm., held at the So. Calif. Edison Co., November 21-22, 1988]. The unit capital costs for the steam-electric and ACC plants are EPRI estimate (EPRI, TAG 1986); that for STIG units is a Bechtel estimate [Soroka, G.E. (Bechtel Eastern Power Corp.). Sept. 1987. Modular remotely operated fully steam-injected plant for utility application. Paper presented at the ASME Cogen-Turbo Conf. Montreux, Switzerland: Am. Soc. Mech. Eng.]; that for the ISTIG is based on estimates made by GE and the staff of the California Energy Commission (M.W. Homer, 1988; Calif. Energy Comm. Staff. 1988. Resource case analysis report. Testimony presented at the Committee Hearing for the 1988 Electricity Report of the Calif. Energy Comm., held at the So. Calif. Edison Co., November 21-22, 1988). The O&M costs are EPRI estimates (EPRI, TAG 1986) for all but STIG and ISTIG units. For the latter the values are those estimated by EPRI for combined cycles (EPRI, TAG 1986), even though a Bechtel analysis indicates that steam-injected gas turbine systems offer inherent O&M cost savings compared to combined cycle units (G.E. Soroka, 1987).

[h] The average gas price for U.S. electric utilities was $2.10/GJ in 1987.

143

Many utility managers are reluctant to consider aeroderivatives in capacity expansion plans. One concern is that, because in their manufacture emphasis is given to the use of special materials to meet the low weight and compactness requirements of jet engines, aeroderivative engines are inherently more costly per kW than industrial turbines, where such constraints are not relevant. While the use of more costly materials does tend to raise the cost of aeroderivatives, a compensating factor is that a greater proportion of the aeroderivative power plant can be built at the factory, where costs are easier to control than in the field. Moreover, the various cycle modifications that would be employed for stationary applications of aeroderivatives tend to lower unit costs. For example, when a simple cycle gas turbine is modified for both steam injection and intercooling, its output can be tripled, resulting in a lower unit capital cost than that of a combined cycle based on an industrial turbine (Table 5.2).[34]

Another concern is that because aeroderivative engines are more delicate than heavy-duty industrial turbines, they are less reliable. This might be true if aeroderivative turbines were maintained like industrial units; instead they are maintained like jet engines. Their compact, modular construction makes it easy to remove and replace failed parts quickly.[35] In fact, the entire basic engine can be removed and replaced with a spare (flown in, if necessary) from a lease-engine pool, resulting in short downtime.[36] With aeroderivative units, it is not necessary to schedule downtime for major maintenance, as is done with heavy-duty industrial units. Also, statistical data on utility use of industrial turbines, combined cycles, and aeroderivative turbines compiled by the North American Electric Reliability Council shows no significant differences in the availabilities of the three types of engines.[37]

A closely related concern is the cost of maintenance. It is widely believed that maintenance costs of gas turbines, heavy-duty industrial as well as aeroderivative, are much higher than those of steam-electric plants. Indeed, between 1982 and 1985 maintenance costs for utility gas turbines averaged 0.76 cents per kWh, compared to 0.26 cents per kWh for coal-fired steam plants.[38] Some utilities report maintenance costs for gas turbines as high as 1.0 to 1.5 cents per kWh.[39] These statistics should be interpreted with care, though, because the data for coal-fired plants are for carefully maintained baseload units, while the gas turbine data are for peaking plants that typically operate at an average capacity factor of only 5 to 7% and are often not carefully maintained. In considering gas turbines for baseload or load-following utility service, a more appropriate historical record is that for gas turbines operated in baseload cogeneration configurations at industrial plants. Preventive maintenance programs carried out over the last 20 years for aeroderivative gas turbines used for cogeneration at the Dow Chemical Company resulted in maintenance costs of 0.2 to 0.3 cents per kWh.[40]

Another concern often expressed about aeroderivative turbines is that utilities will not be interested in them because of their small unit capacities. However, pressed by the financial risks of building large power plants, many utilities are already beginning to shift the focus of their planning efforts to smaller units. Moreover, utilities would be able to improve overall reliability with multiple small units on the same site. The ongoing trend toward more competition in power generation is also making market conditions more favorable for introducing these smaller-scale power-generating technologies.

Steam-Injected Gas Turbines

The most significant development to date relating to stationary power applications of aeroderivative gas turbines was the introduction in the early 1980s of the steam-injected gas turbine (STIG), a variant of the simple gas turbine in which high pressure steam recovered in the HRSG is injected into the combustor, where it is heated to the turbine inlet temperature and then expanded in the turbine (Figure 5.8e).[41] Steam injection can give rise to large increases in power output and electrical efficiency. The only extra work required with steam injection, compared to a simple cycle gas turbine, is that needed to pump the feedwater to boiler pressure, which is negligible compared to the work required to compress the main flow air. This and the fact that the specific heat of steam is double that of air account for the large increases in efficiency and power output that arise with steam injection.[42,43] Aeroderivative engines are chosen for steam injection, because, unlike heavy-duty industrial engines, these units are designed to accommodate mass flows considerably in excess of their nominal ratings, so that only minor modifications are required to operate them as baseload units.[44]

Injecting small amounts of steam (or water) in stationary gas turbines (heavy-duty industrial as well as aeroderivative) for the control of NOx emissions is a well-established practice.[45,46] Only recently has injecting large amounts of steam attracted serious commercial interest as a means of increasing efficiency and power output in stationary applications. Yet the concept is not new. The idea of using steam injection to increase power and efficiency is discussed in textbooks,[47,48] in various articles dating from the mid-1970s,[49-56] and in a 1951 Swedish patent application[57] that was rejected in 1953. The injection of water into gas turbines dates to the earliest use of jet engines, when water was often injected to increase thrust during takeoff.

145

STIG for Cogeneration

The commercialization of STIG for cogeneration applications grew out of the post-PURPA flurry of interest in gas turbine cogeneration in the US The STIG concept was introduced to cope with the most troublesome problem for simple cycle gas turbines in cogeneration applications, that of their poor part-load performance. With a STIG unit, steam not needed for process applications can be injected back into the combustor to produce more electric power. The provisions of PURPA often make it attractive to sell this extra power to the utility, thus extending the economic viability of gas turbine cogeneration to a wide range of variable-load applications.[58] [A combined cycle with a condensing steam turbine using steam extraction to provide steam for process (Figure 5.8d) can also be used economically in variable steam-load applications: steam not needed for process is expanded through the lower turbine stages and condensed to produce more power. But the scale economies associated with steam turbines limit the economical use of the combined cycle to relatively large installations. STIG technology allows the gas turbine to be used in small- scale, variable steam-load applications.]

The first commercially operated STIG cogeneration units involved the use of the Cheng cycle, a patented version of STIG introduced by International Power Technology, Inc.[59,60] Cheng cycle units have been marketed using the Detroit Diesel Allison 501-KH turbine. Without steam injection, this turbine is rated to produce about 3.5 MW of electric power at 24% efficiency when producing power only. With full steam injection, it will produce about 6 MW at 34% efficiency.[61] As of mid-1989, six units based on the Allison 501-KH had been installed and two more ordered; three larger STIG units based on General Electric's LM-5000 had been installed at industrial sites [the first involving an in-the-field modification of a simple cycle cogeneration unit[62]] and fourteen more were either under construction or on order; and seven STIG units based on GE's LM-2500 were being planned (personal communication from M. Horner, Marine and Industrial Turbine Division, General Electric Company, Cincinnati, Ohio, December 13, 1988, and William Flye, Stewart & Stevenson Services, Inc., Houston, December 7, 1988).

The LM-5000, derived from the CF6-50 high-bypass-ratio turbofan engine used in wide-body commercial airplanes (e.g., the DC-10 Series 30, the Boeing 747, and the Airbus A300), is a 33.1 MW unit with a compression ratio of 25:1 and an efficiency of 33% when operated as a simple cycle on natural gas. With full steam injection the output and efficiency of the LM-5000 increase to 51.4 MW and 40% respectively.[63] The LM-2500 is a 21.4 MW simple cycle unit with a compression ratio of

18.5:1 and an efficiency of 33%; with full steam injection it has an output and an efficiency of 26.8 MW and 36%, respectively.[64]

Advanced STIG Cycles for
Central-Station Power

The use of steam injection for cogeneration has stimulated interest for central station applications, in which all the steam raised in the HRSG is injected for power and efficiency augmentation. A paper by a Bechtel analyst indicates that STIG plants based on the LM-5000 and using once-through steam generators would have several advantages over combined cycle units with cooling towers -- including a unit capital cost lower by one-sixth, water requirements less by one-third, a 6% higher availability, and the possibility of remote operation without operators in continuous attendance.[65] A major drawback of STIG is that it is less energy-efficient than combined cycle technology now on the market.[66] Accordingly, despite a modest capital cost advantage for STIG, the busbar cost would usually be lower for combined cycles (Table 5.2).

A more interesting candidate for central-station applications is a proposed modified STIG using intercooling between the two compressor stages (Figure 5.8f).[67] One result of intercooling is that less power is needed to run the compressor. The addition of an intercooler to a simple gas turbine increases the power output but decreases the efficiency; the reduced compressor work requirements would be more than offset by the extra fuel requirements for heating the cooled air exiting the compressor up to the turbine inlet temperature. But modern aeroderivative turbines use air bled from the high-pressure compressor to cool the turbine blades, so that intercooling leads to an efficiency gain as well. Because of the lower temperature of the air used to cool the blades, the metal temperatures can be kept acceptably low, while the turbine inlet temperature is raised. Detailed design work carried out at General Electric indicates that an intercooled STIG (ISTIG) based on the LM-8000[68] will be able to operate at a turbine inlet temperature of 1370°C and produce about 114 MW with an average efficiency of 48.3% and a guaranteed efficiency of 47.0%, at an installed capital cost of $400 per kW.[69]

The projected ISTIG efficiency is somewhat higher than that for an advanced combined cycle and its estimated capital cost is somewhat less, leading to a lower busbar cost (Table 5.2). The busbar cost would probably be less than for a large coal-fired steam-electric plant with flue gas desulfurization even if the natural gas price is double the average for 1986 (Table 5.2).

The indicated efficiency advantage of the ISTIG compared to the combined cycle is not the result of a systematic comparison of

147

steam-injected and combined cycle designs. Moreover, the estimated performance difference is too small to declare unequivocally that steam-injected designs are more efficient. In looking to the future, the balance could tip in favor of combined cycles, for example, if the Kalina cycle were successfully developed and used instead of the steam Rankine cycle in combined cycles. But there are also many possible modifications to the ISTIG cycle.

One such cycle modification involves reheat, or the addition of an additional combustor after the higher pressure turbine stages and before the power turbine. (The power turbine generates the net cycle power; the output of the higher pressure turbine stages drives the compressor.) Since combustion in the gas turbine takes place with a large amount of excess air (needed to keep the combustion product gases sufficiently cool that the metallurgical temperature limits on the turbine blades are not exceeded), there is oxygen available to burn more fuel in a reheat combustor. If a reheat combustor is added to an ISTIG unit based on General Electric's LM-8000, the power output would increase to 180 MW and the efficiency to 52%.[70] The cycle efficiency would increase because the average temperature at which heat is added to the cycle would thereby increase, while the temperature of the heat discharged to the atmosphere would remain the same.

With reheat, not only do the power output and efficiency increase, but also the temperature of the exhaust gases from the power turbine increases enough that it becomes feasible to use some of the turbine exhaust heat to reform the fuel with steam in the presence of an appropriate catalyst.[71,72,73,74] When methane fuel is reacted with steam in the reformer, some of the methane is converted into a mixture of hydrogen, carbon monoxide, and carbon dioxide. As the steam-reforming reaction is highly endothermic, the chemical energy content of the products is greater than that of the fuel from which it is derived; thus through steam reforming, low-quality heat can be converted into high-quality chemical energy. To the extent that some of the turbine exhaust heat can be used for chemical recuperation as an alternative to heat recuperation through steam injection, there would be a net cycle efficiency improvement because of the reduction of the latent heat loss to the stack. (More than half the heat used to raise steam in the HRSG is the latent heat needed to evaporate water, which is lost to the stack in a STIG cycle.) It is estimated that the addition of a steam reformer to an LM-8000-based, natural gas-fired ISTIG unit with reheat increase the efficiency to 54% but reduce its output to about 160 MW.[75]

Thus, the performance of both combined cycles and advanced STIG units could be improved, even without further improvements in turbine inlet temperature or turbine blade cooling technology. Because of the uncertainties relating to an efficiency comparison of advanced STIG and combined cycle designs, decisions to commercialize advanced STIG technologies should be made on grounds other than efficiency alone.

An ISTIG has several advantages over a combined cycle unit: it is simpler, requiring no steam turbine, condenser, or cooling tower; pollution controls would be less costly than with combined cycle units; the small unit capacities of ISTIG units implies flexibility in capacity planning, improved reliability, and ease of maintenance through lease-pool arrangements; their small size also makes them good candidates for cost-cutting innovations and the economies of mass production; and steam-injected gas turbines will continue to benefit quickly from expected continuing improvements in jet engine technology.

A drawback of STIG cycles is that significant quantities of steam are exhausted to the atmosphere. While the absolute level of makeup water requirements actually favors steam-injected cycles, some 0.6 kg per kWh for an ISTIG unit, compared to 4.1, 1.8, and 0.8 kg per kWh, for a nuclear plant, a large fossil-fuel fired steam-electric plant, and an advanced combined cycle, respectively,[76] all the water required for STIG cycles must be processed to boiler quality. In fact, STIG cycles require about five times the high-quality water and demineralization processing capacity required for combined cycles.[77] Although these water processing requirements typically would not give rise to significant economic penalties,[78] it could be a source of concern in water-scarce situations. But in such instances makeup water requirements could be reduced to zero by condensing water vapor out of the exhaust stream of the heat recovery steam generator for reuse, using a condensing heat exchanger. For STIG units, it has been estimated that complete water recovery could be achieved for an 11% increase in capital cost and a 1.2% increase in fuel requirements per kWh.[79,80]

Such considerations, collectively considered, suggest that advanced STIG cycles warrant development. It has been estimated that to develop ISTIG would take four to five years and cost $100 million, including the cost ($40 million) for the first unit.[81] As no proof-of-concept is involved, only good engineering design, the technological risk associated with development is small. Accordingly, bringing the technology to market requires only the sale of a few units to pay for the relatively modest development costs.

149

Gas Turbines and Long-Term
Natural Gas Supply Considerations

Gas turbines have not been exploited much for power generation in part because of concerns by some energy planners about the long-term availability of natural gas. But such concerns are becoming less and less important in decisions relating to power generation, for several reasons.

First, the natural gas supply outlook is now quite favorable for the decades immediately ahead -- especially in many developing countries, but even in the United States[82] and Western Europe,[83] where concerns about natural gas supply have led to public policies restricting the use of natural gas for power generation.

Second, the prospect of a continuing stream of innovations in gas-turbine power-generating technology while steam-electric technology stagnates has prompted massive investments in research and development aimed at marrying coal to the gas turbine through the use of integrated coal gasifier/gas turbine power systems, in the United States, Western Europe, and Japan.[84,85] These efforts led to a commercial demonstration project proving the technology in the 100 MW integrated coal gasifier/combined cycle power plant at Cool Water, California, and to extensive follow-on developmental efforts that hold forth the promise that second-generation coal-gasifier/gas turbine power generating technology is likely to be both cleaner and less costly than coal-fired steam-electric power generation with flue gas desulfurization.[86]

The future of coal gasifier/gas turbine power generation is uncertain because technologies for efficiently removing sulfur are not yet commercially proven,[87] because low natural gas prices make it hard for coal gasifier/gas turbines to compete, and because growing concerns about the greenhouse warming may lead to constraints on coal use in many parts of the world. But if coal is to have a significant future in power generation, the coal-based technologies of choice will probably be configurations involving integrated coal gasifier/gas turbine systems, because they offer both higher efficiencies (and thus lower CO_2 emissions per kWh) and lower unit capital costs than alternative coal-based power generation technologies.[88] In any case, a prudent "evolutionary" strategy for introducing coal-fired gas turbines would involve widespread initial use of natural gas-fired gas turbines in power generation. This strategy would help create a dynamic, rapidly growing gas-turbine manufacturing industry that would be a favorable theater for innovation, and it would provide utilities with a broad base of experience with gas turbine technology before shifting to more complicated coal-based gas-turbine systems. Various strategies have been advanced in recent years that would enable utilities to

150

start up their gas turbine systems with natural gas, with the flexibility of shifting later to coal, if natural gas prices rise too much.[89,90]

Third, the power generation options for the long term are much broader than was once thought. Most of the developments relating to coal gasifier/gas turbine technology are readily adaptable to firing with biomass. STIG and ISTIG technologies, in particular, offer relatively low unit capital costs and high thermodynamic efficiencies at the modest sizes (less than 100 MW) needed for biomass applications. The prospects are good for bringing to market in the early 1990s biomass-fired versions of such turbines that will be able to compete with both coal and nuclear power technologies in many circumstances.[91,92] Biomass-fired gas turbines will be important in many parts of the world not only because biomass resources are more widely available than natural gas and coal resources, but also because they offer a strategy for helping to cope with the problem of global warming. If the biomass is grown renewably, its use results in no net buildup of carbon dioxide in the atmosphere. Moreover, if the biomass is grown in plantations on previously deforested or unforested land, the buildup of the biomass inventory will extract CO_2 from the atmosphere, and the steady-state inventory will be a reservoir of sequestered carbon.

It is also very likely that solar photovoltaic (PV) technology will begin to be used in power-generating applications, beginning in the 1990s, as rapid progress is being made in both high-efficiency, concentrating crystalline solar cell and low-cost, thin-film technologies.[93,94]

All of these future developments mean that emphasis on low capital-cost, energy-efficient, gas turbines fired with natural gas will make good sense for power generation in most parts of the world in the decades immediately ahead, regardless of the long-term outlook for natural gas supplies and natural gas prices. Natural gas is likely to be the fossil fuel of choice during the transition to the post-fossil fuel era, and during this transition, gas turbine-based power generation is likely to be one of the most important markets of natural gas.

Environmental Aspects of
Advanced Gas Turbine Technologies

Natural gas-fired gas turbines emit negligible amounts of sulfur oxides and particulates. The high combustion temperatures lead to high emissions of nitrogen oxides (NO_x), however. Their uncontrolled NO_x emissions (Table 5.3) exceed the levels established in the federal New Source Performance Standards (NSPS) for natural gas-fired gas turbines

151

Table 5.3. Estimated NO_x Emissions for Alternative Cycles Based on the LM-5000[a]

	Adiabatic Flame Temperature ($T^\circ K$)		NO_x Emission Rate (ppm @ 15% O_2)	
Uncontrolled Simple Cycles[b]	2530		260	
Steam-Injected Cycles	STIG	ISTIG	STIG	ISTIG
Dry[c]	2548	2471	301	159
S/Fc = 1.0	2406	2329	89.5	43.6
S/Fc = 2.0	2274	2199	25.3	11.6
S/Fc = 3.0	2158	2085	7.4	3.2
Chemically Recuperated, Steam-Injected Cycles	CRSTIG	CRISTIG	CRSTIG	CRISTIG
Dry[c]	2223	2161	15	7.6
S/F[c] = 0.2	2107	2048	4.1	2.0
S/F[c] = 0.3	2055	1997	2.2	1.0
S/F[c] = 0.5	1959	1905	0.63	0.29

[a] The adiabatic flame temperature (AFT) is calculated for stoichiometric conditions -- for methane @ $300^\circ K$ in simple and dry STIG/ISTIG cycles; for methane @ $400^\circ K$ in wet STIG/ISTIG cycles; for a mixture (@ $950^\circ K$) of 10% methane, 62% steam, 22% hydrogen, 1% carbon monoxide, and 5% carbon dioxide (a typical fuel composition expected from the steam-reforming of natural gas) for CRSTIG/CRISTIG cycles. (A reheat turbine would be needed to achieve the turbine exhaust temperature needed for steam reforming. Because of the reduced oxygen level in the reheat combustion "air," the AFT and hence NO_x formation in the reheat combustor would be significantly lower than in the primary combustor. This effect is neglected here.) The temperature of steam for NO_x control is assumed to be $578^\circ K$. For the calculated AFT, the indicated NO_x emission level is obtained from an empirical relationship between AFT and NOx emissions (Sidebotham, G.S., Williams, R.H. 1989. Preliminary report on NO_x and cogeneration in New Jersey. Princeton, N.J.: Cent. Energy and Environ. Stud., Princeton Univ.) For the lowest AFTs considered, empirical data are not yet available, and the indicated NO_x emissions, obtained by extrapolation, may underestimate actual emissions. While thermal NO_x emissions are strongly correlated with AFT, prompt NO_x emissions are not. At high AFT, thermal emissions dominate; at low AFT, prompt NO_x emissions become more important. For a compression ratio of 27.6:1, 31.3:1, and 33.5:1, and an air temperature at the compressor exit of $787^\circ K$, $816^\circ K$, and $658^\circ K$, for the simple cycle, STIG/CRSTIG cycles, and ISTIG/CRISTIG cycles, respectively.

[b] The uncontrolled NO_x emissions rate for this simple cycle turbine is higher than for the turbines used in combined cycles, because the low pressure ratio of the gas turbines used in combined cycles leads to a lower compressor exit temperature and hence a lower AFT.

[c] "Dry" means that all steam is injected with the dilution air (i.e. no steam is injected into the primary combustion zone). S/F is the molar ratio of steam/fuel for steam injected into the primary combustion zone for NO_x control.

promulgated in the US in 1977[95] and are far in excess of the standards in some states with especially severe air quality problems.[96]

Among available technologies for reducing NO_x emissions, a well-established approach involves the injection of steam or water into the primary combustion zone. NO_x emissions tend to fall exponentially with the ratio of steam or water injected in the primary combustion zone to fuel, as demonstrated empirically for steam in the first field-modified STIG unit derived from the LM-5000 (Figure 5.11). NO_x is controlled when steam or water is injected into the primary combustion zone because so doing reduces the flame temperature (Table 5.3).

With water injection, the percentage increase in fuel required to bring the mixture up to the turbine inlet temperature exceeds the percentage increase in power output resulting from the increased turbine mass flow, so that the electrical efficiency is reduced. With steam injection, the electrical efficiency is also reduced for combined cycle systems (where the optimal use of the steam produced in the HRSG is for power generation in the steam turbine),[97] but not for STIG or ISTIG cycles, where NO_x control is an automatic benefit.

Dramatic reductions in NO_x emissions are also possible with chemically recuperated gas turbine cycles, with and without additional steam injection for NO_x control, as indicated for chemically recuperated STIG and ISTIG (CRSTIG and CRISTIG) cycles in Table 5.3. As with steam and water injection, the chemically recuperated cycles achieve low NO_x emissions levels because with steam-reformed fuel, flame temperatures are lower than with natural gas fuel (since the heating value of the steam-reformed fuel is much lower *per mole* than that of the fuel from which it is derived). CRSTIG and CRISTIG cycles are also more efficient electricity generators than cycles that do not use steam-reformed fuel.[98] The combination of low NO_x emissions and high efficiency achievable with these cycles has led the South Coast Air Quality Management District (SCAQMD) in California and the California Energy Commission to examine chemically recuperated gas turbine cycles as a promising technological strategy for reducing NO_x emissions in Southern California while simultaneously controlling costs (personal communication from Jack Janes, California Energy Commission, December 15, 1988).

CO emissions increase as the level of steam injection increases, while NO_x emissions are declining. Figure 5.12 shows, for example, that if NO_x is controlled in a natural gas-fired LM-5000 to levels below 25 ppm, the CO emissions would be in excess of 50 ppm. Determining the optimal level of steam injection into the primary combustion zone involves balancing considerations of both NO_x and CO emissions. If, by the time

Figure 5.11. NOₓ Reduction Ratio as a Function of Ratio of Weight of Steam Injected into Primary Combustion Zone to Weight of Fuel*

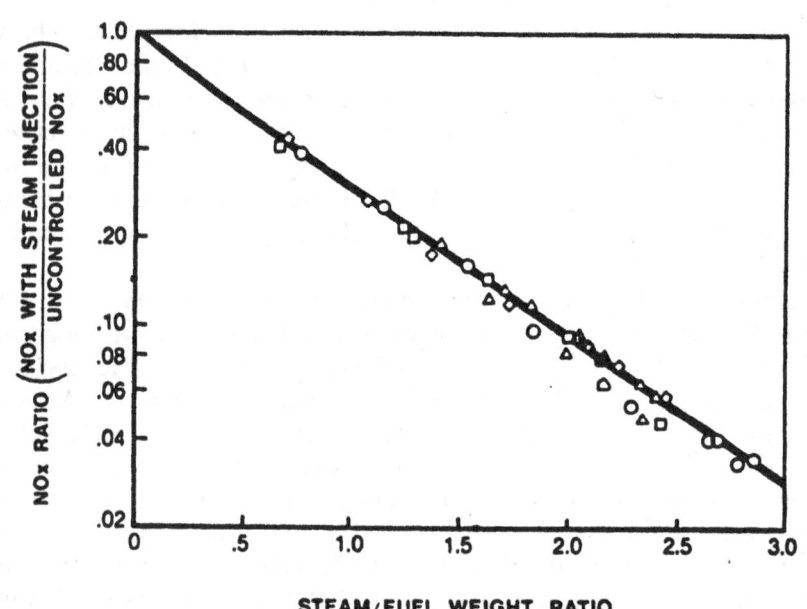

*Measured at the First Field-Modified STIG Unit Based on the LM-5000, at Simpson Paper Co., Anderson, Calif.

Source: Burnham, J.B., Giuliani, M.H., Moeller, D.J. June 8-12, 1986. Development, installation, and operating results of a steam injection system (STIG) in a General Electric LM 5000 gas generator. Paper 86-GT-231. Presented at the Int. Gas Turbine Conf. and Exhib., Dusseldorf, West Germany: Am. Soc. Mech. Eng.

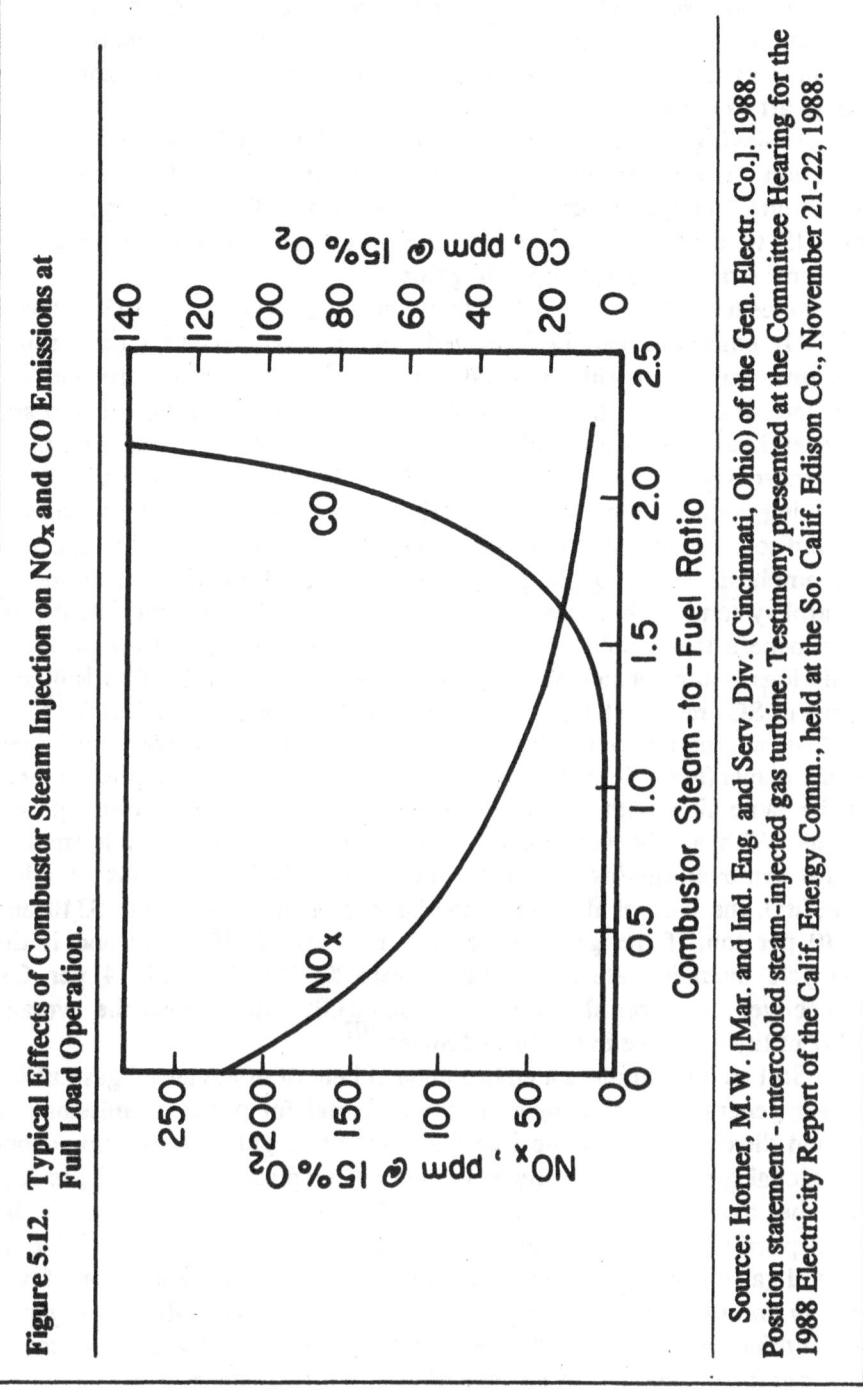

Figure 5.12. Typical Effects of Combustor Steam Injection on NO$_x$ and CO Emissions at Full Load Operation.

Source: Horner, M.W. [Mar. and Ind. Eng. and Serv. Div. (Cincinnati, Ohio) of the Gen. Electr. Co.]. 1988. Position statement - intercooled steam-injected gas turbine. Testimony presented at the Committee Hearing for the 1988 Electricity Report of the Calif. Energy Comm., held at the So. Calif. Edison Co., November 21-22, 1988.

the maximum acceptable CO emissions level is reached, still higher levels of steam injection are desired for power and efficiency augmentation, extra steam is injected sufficiently far from the primary combustion zone so as not to affect pollutant emissions further.

Achieving extremely low levels of NO_x emissions with steam injection might require supplemental control of carbon monoxide emissions. Catalytic oxidation of the exhaust gas (the technology used in the US to control CO emissions from cars) is a promising, relatively low-cost strategy for accomplishing this.[99]

Since natural gas contains virtually no sulfur and since very low levels of NO_x emissions can be achieved with natural gas-fired gas turbine systems, natural gas-fired gas turbines can also be used to cope with the problem of acid rain. In the United States, the large quantities of SO_2 and NO_x emitted by existing coal-fired steam-electric power plants are among the most significant pollutant emissions leading to acid rain (100). Growing concern about acid rain will probably lead to US legislation in the not-too-distant future aimed at curbing these emissions. While much of the emphasis in the ongoing debates is on requiring the retrofitting of control systems such as stack gas scrubbers on these old power plants, the only benefit that would be gained thereby is a reduction of the pollutant emissions -- at a considerable cost, estimated to be some \$340 to \$640 per tonne of SO_2 removed, for an average sulfur content in coal of 3.1%.[101]

An alternative would be to encourage scrapping existing coal-fired steam plants (regardless of their remaining useful lives) and replacing them at the same sites with new advanced natural gas-fired gas turbine power plants. With all the incremental costs of the new facilities allocated to sulfur removal (giving no credit for reduced NO_x emissions or other benefits), the cost of this scrap-and-build strategy would cost \$340 and \$640 per ton, if the gas turbine systems were ISTIG units and if the lifecycle average natural gas price were \$3 per GJ and \$4 per GJ, respectively -- prices that were 50% and 100% higher than the average 1987 utility gas price in the United States.[102]

Still another important environmental benefit of advanced gas turbine power generating technologies is the potential for reduced emissions of carbon dioxide, which is desirable in light of heightened concerns about the global greenhouse warming. In general, emphasis on natural gas fuel for the transition to the post-fossil fuel era would help slow the atmospheric build-up of carbon dioxide. Burning one energy unit of natural gas releases just 0.55 times as much CO_2 as the combustion of one energy unit of coal. Furthermore, generating electricity with natural gas in gas turbines as efficient as ISTIG would release just 0.4 times as much CO_2 per kWh generated as a conventional coal-fired steam plant.

156

Potential Applications of
Advanced Gas Turbines

Natural Gas Resources

Natural gas will dominate initial applications of advanced gas turbines, because gas supplies will be abundant in many parts of the world in the decades immediately ahead. According to estimates by the US Geological Survey, there is about as much conventional natural gas left in the world as conventional crude oil, but globally gas is used at just half the rate oil is (Table 5.4; see also Chapter 1). Remaining gas resources for the US and Canada are about 50% greater than oil resources, while for all industrialized countries remaining gas resources are more than twice as large as remaining oil resources (Table 5.4).

The outlook for gas is especially promising for developing countries. Natural gas resources exist in about 50 developing countries, including 30 that import oil.[103] Also, gas resources are large in relation to gas production in developing countries; although they have about as much gas as industrialized countries, they produce it at just one-fifth the rate it is produced in industrialized countries. Outside the Middle East, developing country gas resources are equivalent to more than a 200-year supply at the current rate of production (Table 5.4).

US Applications

Even in the US, where natural gas resources are more fully developed than in most of the rest of the world, natural gas-fired gas turbines can play important roles in power generation in the decades immediately ahead.

One important initial application for ISTIG technology would involve replacing the 127 GW of existing oil and gas-fired steam-electric plants expected to be operating in the US in 2000[104] with natural gas-fired ISTIG units. While these are load-following steam plants typically operated at low capacity factor [projected by DOE to average 46% in 2000[105]], they are so inefficient (32%) that it would be worthwhile replacing them with ISTIG units, even with fuel prices as low as $2 per GJ ($12 per barrel of oil equivalent).[106] Doing so for all oil and gas-fired steam plants in the US would lead to producing the same amount of electricity as DOE projected for such plants in 2000[107] while saving the fuel equivalent to 0.82 million barrels per day of oil. Considerations of both the cost savings potential and the large NO_x reduction potential (especially important in Southern California) of ISTIG technology, led the staff of the California Energy Commision to a recommmendation along these lines -- that Southern California Edison Company retire some 5700

Table 5.4. Natural Gas and Oil Resources and Production

	Natural Gas Resources[a] (EJ)	Natural Gas Production[b] (EJ/Year)	Crude Oil Resources[a] (EJ)	Crude Oil Production[b] (EJ/Year)
Industrialized Countries				
US/Canada	1079.0[c]	21.0	720.2	23.3
Western Europe	481.6	7.3	318.8	8.6
Australia/New Zealand	149.3	0.6	36.7	1.3
USSR/Eastern Europe	2757.6	27.1	991.9	25.9
Subtotal	4467.5	56.0	2067.6	59.1
Developing Countries				
Central America	228.3	1.3	342.7	6.5
South America	290.4	1.7	523.8	7.7
Asia	779.5	3.9	641.3	11.6
Africa	620.9	2.0	627.2	11.1
Middle East	2316.8	2.6	3175.1	23.0
Subtotal	4235.9	11.5	5310.1	59.9
Global Total	8703.4	67.5	7377.7	119.0

[a] Proved reserves plus estimated reserve appreciation in discovered fields plus estimated recoverable undiscovered resources, as of January 1, 1985 (Masters, C.D., Attanasi, E.D., Dietzman, W.D., Meyer, R.F., Mitchell, R.W., Root, D.H. 1987. *World Resources of Crude Oil, Natural Gas, Natural Bitumen, and Shale Oil.* Paper prepared for the 12th World Petrol. Cong. Houston, Tex.).

[b] Production in 1985 [Energy Inf. Admin. 1987. *International Energy Annual 1986.* DOE/EIA-0219(86). Washington, D.C.: US GPO].

[c] The resource estimate given here for the US (650 EJ) may underestimate remaining gas resources. A more recent assessment carried out for the US Department of Energy under the auspices of the Argonne National Laboratory estimated that remaining gas resources in the lower 48 states of the US. recoverable at wellhead prices less than $2.75/GJ amount to 633 EJ, with 189 EJ of additional resources recoverable at costs in the range $2.75/GJ to $4.60/GJ (Argonne National Laboratory. May 1988. An assessment of the natural gas resource base of the United States. Washington, D.C.: Off. Policy, Planning, and Analysis, US Dept. Energy).

MW of existing gas-fired steam-generating capacity in favor of ISTIG plants.[108]

Developing Country Applications

While the abundance of their natural gas resources (Table 5.4; see also Chapter 1) suggests that natural gas could play a major role in the energy economies of many developing countries, the use of natural gas is presently inhibited by the lack of gas transmission and distribution infrastructure, which is costly to develop. This problem could often be overcome if power generation were emphasized as an initial market. Some of the large revenues generated in power sales could be used to pay for the construction of the gas delivery system, thereby helping to make gas available to other users at reasonable cost.[109]

Its low capital cost makes the gas turbine an especially attractive technology for developing countries in light of the unaffordabiliy of capital investments for electricity based on conventional sources (Table 5.1).

Not only are the overall capital requirements small for these advanced gas turbine power plants, but also, many industrializing countries could draw on indigenous management and engineering talent for much of the design and construction effort required. The power turbine, the heat recovery steam generator, and the electrical generator, as examples, are system components that can be readily manufactured in many parts of the world. The part of the system for which it may be difficult to avoid expenditures of foreign exchange is the "gas generator," the "high technology" part of the system that is derived from a jet engine. The gas generator actually accounts for only a modest fraction of the total power plant cost; the mass-produced CF6 jet engine, from which LM-5000 STIG and ISTIG units would be derived, costs only about $6 million -- thus contributing only $53 per kW to the cost of the ISTIG unit.[110]

The scale characteristics of aeroderivative turbines are also well suited to developing countries. In most developing countries, the total utility grid capacity is too small to be well matched to much larger hydroelectric or steam-electric power plants. Adding new capacity in small increments with gas turbines makes it possible to avoid the alternating periods of power glut and power shortage associated with utility planning based on large plants and can lead to improved system reliability.

The compact, modular nature of aeroderivative turbines makes it possible to replace failed parts and even whole engines quickly with replacements flown or trucked in from centralized maintenance facilities.[111] This feature of the aeroderivatives is especially attractive for many developing countries, where sophisticated maintenance capability is typically unavailable at power generating sites. The required maintenance

159

network is already in place in most developing countries that have their own commercial airlines; their planes are typically maintained through centralized lease-pool arrangements. This advantage is reflected, for example, in the fact that of the 210 General Electric LM-2500 aeroderivative turbines in service throughout the world as of 1986, 54, 8, and 26 were being used in developing countries of Latin America, Africa, and Asia, respectively (personal communication from L. Gelfand, Marine and Industrial Turbine Division, General Electric Company, Cincinnati, Ohio, February 1987).

Conclusion

For the next two to four decades, advanced gas turbines offer multiple benefits for power generation. The prospects of reducing electric power costs in both industrial and developing countries, of reducing local air pollution and acid rain emissions to low levels, of reducing carbon dioxide emissions to levels considerably below those associated with coal-fired steam-electric plants, and of avoiding the risks of expanded dependence on nuclear power, are benefits not easily matched by alternatives.

While the benefits resulting from the wide use of heavy-duty industrial gas turbines would be large, there are good reasons for also bringing advanced aeroderivative turbines into wide use. It appears that controlling pollutant emissions responsible for acid rain to very low levels would be more readily accomplished with various advanced aeroderivative turbines than with combined cycles. Also, with aeroderivatives the advantages of high efficiency and low unit capital costs can be extended to modest scale, resulting in greater flexibility in capacity planning, improved reliability, and ease of maintenance. And the small size of aeroderivatives makes it possible to reverse the trend in power technology toward costly field construction and bring most construction work back to the factory, where the economies of mass production can be exploited. Moreover, aeroderivative turbines will continue to benefit more directly from improvements in jet engine technology than heavy-duty industrial turbines.

Wide use of advanced gas turbines would *not* solve the electric power problem for all time. Eventually, the tightening of world gas supplies and concerns about the atmospheric buildup of carbon dioxide will limit the attractiveness of further expanding the use of these engines for power generation with fossil fuels. However, a major shift to fossil fuel-fired gas turbines for power generation in the decades immediately ahead would buy time to develop alternative clean power sources for the long term.

REFERENCES AND NOTES

1. Hass, J.E., Mitchell, E.J., and Stone, B.K., *Financing the Energy Industry*. Cambridge, Mass., Ballinger, 1974.

2. In this paper conversions to 1987 dollars were made using the US gross national product deflator, if original data were in the nominal dollars of other years.

3. International Energy Agency, *Electricity in IEA Countries: Issues and Outlook*, Paris, 1985.

4. Electric Power Research Institute, *Technical Assessment Guide. 1: Electricity Supply -- 1986*, Palo Alto, Calif, 1986.

5. Hass, et al., op. cit.

6. International Energy Agency, op. cit.

7. Electric Power Research Institute, op. cit.

8. Schneider, H.K., *Investment Requirements of the World Energy Industries 1980-2000*. London, England, World Energy Conference, 1987.

9. The WEC capital expenditure estimates are for electricity demand growth rates for developing countries in the range of 4.5% to 6.8% per year, 1980-2000. For comparison, the growth rate has averaged about 7% since 1980, and the long-term historical average growth rate has been about 9% per year.

10. Fisher, J.C., *Energy Crises in Perspective*, New York, Wiley, 1974.

11. Loose, V.W., and Flaim, T., "Economies of Scale and Reliability: the Economics of Large Versus Small Generating Units." *Energy Systems and Policy*, Vol. No. 4 (1-2), pp. 37-56, 1980.

12. Duffy, T.E., Schneider, P.H., and Campbell, A.H, *Advanced High Performance Steam Systems for Industrial Cogeneration*, Solar Turbines International, Washington, D.C., US Dept. of Energy, 1986.

13. Electric Power Development Co., Ltd., *Wakamatsu 50 MW Demonstration Plant for Fluidized-Bed Combustion Boiler and Ultrasuper Critical Turbine*. Kyushu, Japan, Wakamatsu Coal Utilization Center, 1988.

14. Combustion Eng., Inc., *Combustion: Fossil Power Systems*, Windsor, Conn., Rand McNally, 1981.

15. Fraas, A.P., *Engineering Evaluation of Energy Systems*, New York, McGraw-Hill, 1982.

16. Wolfe, W.R., *Energy Conversion Alternatives Study (ECAS), Westinghouse Phase I Report, vol. XI: Advanced Steam Systems*, NASA, 1976.

17. El-Sayed, Y.M., and Tribus, M., *A Theoretical Comparison of the Rankine and Kalina Cycles*, Cambridge, Mass., Center for Advanced Engineering Study, Massachusetts Institute of Technology, 1986.

18. Mechanical Technology, Inc., *Technology Assessment of Advanced Power Generation Systems II -- Kalina Bottoming Cycle*, Palo Alto, Calif., Electric Power Research Institute, 1986.

19. Rankin, B., "Revolutionary Power Plant to Get Its Day in the Sun, *New Technology Week*, Dec. 21, 1987, p. 5.

20. Mechanical Technology, Inc., op. cit.

21. Gluckman, M.J., *The Future of Gas Turbine-based Power Generation in the United States*, Paper presented at the 1988 ASME COGEN-TURBO II International Symposium & Exposition, Montreux, Switzerland, American Society of Mechanical Engineers, 1988.

22. Electric Power Research Institute, op. cit.

23. Energy Information Administration, *Historical Plant Cost and Annual Production Expenses for Selected Electric Plants 1985*, Washington, D.C., 1987.

24. Fraas, op. cit.

25. Ibid.

26. Energy Information Administration, 1987, op. cit.

27. Bureau of the Census, US Dept. Commerce, *Statistical Abstract of the United States, 1987*. Washington, D.C., 1987.

28. Directorate for Inf. Oper. and Rep. Annual. Prime contractor awards by service category and federal supply classification. Washington, D.C., US Dept. of Defense.

29. A significant difference between the steam turbine and the gas turbine is that the maximum temperatures of the metal and the working fluid are oppositely related for these technologies. In steam turbine systems the maximum temperature of the metal is always higher than that of the working fluid; thus metallurgical contraints that limit the maximum metal temperature make it difficult to increase efficiency by increasing the working fluid temperature. But in gas turbine-based cycles the metal temperature is lower than that of the working fluid, and the difference between these temperatures is increasing continually with improvements in turbine blade cooling technology, leading to continual improvements in efficiency.

30. Energy Information Administration, *Annual Energy Outlook 1987*, with projections to 2000, Washington, D.C.

31. Brandt, D.E., "Heavy-Duty Turbopower: the MS7001F." *Mechanical Engineering*, July, 1986, pp. 28-36.

32. Ibid.

33. "Can Gas Turbines Be Twice as Good?" *Global Gas Turbine News*, April 1987.

34. General Electric's LM-5000 is the 33 MW aeroderivative gas turbine which provided the basis for the first design of a 110 MW intercooled steam-injected gas turbine (ISTIG). [See Engineering Department, Pacific Gas and Electric Co., *Scoping Study: LM5000 Steam-Injected Gas Turbine*. Based on work performed by the Mar. and Ind. Engine Proj. Dept. (Evendale, Ohio) of the General Electric Co., 1984.] The LM-5000 in turn is derived from the CF6 turbofan jet engine which sells for $6 million. (See Ungeheuer, F. 1988, and "They Make Good Things for Flying," *Time*, May 2, p. 55.) As the CF6 weighs 4770 kg, it is indeed costly on a per unit mass basis -- $1260 per kg. (For comparison, at the time of this writing, gold and silver were selling on the world market for $14,500 and $225 per kg, respectively.) But this "high technology" part of ISTIG contributes only $53 per kW or 13% of its estimated installed cost of $400 per kW (Table 5.2).

35. Complete inspection (with any necessary replacements) of the hot section of a GE LM-2500 aeroderivative turbine requires a crew of five working 100 person-hours. [See Jackson, J., *Aircraft-Derivative Gas Turbine Maintenance Practices*, Schenectady, N.Y., General Electric Gas Turbine Division, 1984.] The same job requires a six-person crew working 480 person-hours for a GE Series 5000 industrial turbine that has a comparable output. [See Knorr, R., *Heavy-Duty Gas Turbine Maintenance Practices, Schenectady, N.Y., General Electric Gas Turbine Division, 1984.]*

36. This possibility arises because the gas generator, the "high technology" part of an aeroderivative engine, for which maintenance is most crucial, is easily transported. The gas generator for the largest aeroderivative turbine available, General Electric's LM-5000, weighs just 4770 kilograms and measures only 1.8 m x 2.1 m x 4.6 m.

37. N.J. Energy Conservation Laboratory, *Report on the NJECL Workshop on Steam-Injected Gas Turbines for Central Station Power Generation, eds. Larson, E.D., and Williams, R.H., Princeton, N.J., Center for Energy and Environmental Studies, Princeton University, 1986.*

38. Energy Information Administration, 1987, op. cit.

39. N.J. Energy Conservation Laboratory, op. cit.

40. Ibid.

41. The injected water must be treated to avoid turbine blade corrosion problems. Because the minimum water treatment level required is not yet known (N.J. Energy Conserv. Lab., op. cit.), present practice is to be conservative. Even so, water treatment costs are minor. For STIG cogeneration units based on the Allison 501-KH, water treatment costs

have been estimated to be 0.09 cents per kWh (for 1.63 liters per kWh and water treatment costs of 0.05 cents per liter, personal communication from Koloseus, C., International Power Technology, Inc., April 1985). For central- station STIG units based on the GE LM-5000, the capital cost for make-up and waste-water treatment (based on typical river water quality in the Eastern US) has been estimated to be less than $20 per kW, some 5% of the total installed cost. (See Soroka, G.E., Bechtel Eastern Power Corp., Modular Remotely Operated Fully Steam-Injected Plant for Utility Application, Paper presented at the ASME Cogen-Turbo Conf. Montreux, Switzerland, American Society of Mechanical Engineers, Sept. 1987.

42. Larson, E.D., and Williams, R.H., "Steam-Injected Gas Turbines," *ASME J. Eng. for Gas Turbines and Power*, Vol. 109, No. 1, 1987, pp. 55-63.

43. Larson, E.D., and Williams, R.H., *A Primer on the Thermodynamics and Economics of Steam-Injected Gas-Turbine Cogeneration*, PU/CEES Report No. 192. Princeton, N.J.: Center for Energy and Environmental Studies, Princeton University, 1985.

44. Leibowitz, H., and Tabb, E., "The Integrated Approach to a Gas Turbine Topping Cycle Cogeneration System," *ASME J. Eng. Gas Turbines and Power*, Vol. 106, 1984, pp. 731-36.

45. Allen, R., and Kovacik, J., "Gas Turbine Cogeneration -- Principles and Practice," *ASME J. Eng. Gas Turbines and Power*, Vol 106, 1984, pp. 725-30.

46. Touchton, G., "Influence of Gas Turbine Combustor Design and Operating Parameters on Effectiveness of NOx Suppression by Injected Steam or Water." *ASME J. Eng. Gas Turbines and Power*, Vol. 107, 1985, pp. 707-13.

47. Diamant, R.M.E., *Total Energy*, New York, Pergamon, 1970.

48. Hayward, R.W., *Analysis of Engineering Cycles*, New York, Pergamon, 1980.

49. Leibowitz and Tabb, op. cit.

50. Maslennikov, V., and Shterenberg, V.Y., "Power Generating Steam Turbine-Gas Turbine Plant for Covering Peak Loads," *Thermal Eng*, Vol. 21, No. 4, 1974, pp. 82-88.

51. Fraize, W.E., and Kinney, C., "Effects of Steam Injection on the Performance of Gas Turbine Power Cycles," *J. Eng. Power*, Vol 101, 1979, pp. 217-27.

52. Davis, F., and Fraize, W., *Steam-Injected Coal-Fired Gas Turbine Power Cycles*, MTR-79W00208. McLean, Va., Mitre Corp., 1979.

53. Brown, D.H., and Cohn, A., "An Evaluation of Steam-Injected Combustion Turbine Systems." *J. Eng. Power*, Vol. 103, 1981, pp. 13-19.

54. Kosla, L., Hamill, J., and Strothers, J., "Inject Steam in a Gas Turbine -- But Not Just for NO$_X$ Control." *Power*, Feb. 1983.

55. Digumarthi, R., and Chang, C.N. 1984. "Cheng-Cycle Implementation on a Small Gas Turbine Engine," *J. Eng. Gas Turbines and Power*, Vol. 106, 1984, pp. 699-702.

56. Jones, J.L., Flynn, B.R., and Strother, J.R., *Operating Flexibility and Economic Benefits of a Dual-Fluid Cycle 501-kb Gas Turbine Engine in Cogeneration Applications*, Paper 82-GT-298. N.Y., American Society of Mechanical Engineers, 1984.

57. Nicolin, C., "A Gas Turbine with Steam Injection," *Swedish Patent Application No. 8112/51*, Stockholm, 1951.

58. Larson and Williams, 1987, op. cit.

59. Cheng, D.Y., "Regenerative Parallel Compound Dual-Fluid Heat Engine," *US Patent No. 4,128,994*. Washington, D.C., 1978.

60. Cheng, D.Y., "Control System for Cheng Dual-Fluid Cycle Engine System," *US Patent No. 4,297,841*. Washington, D.C., 1981.

61. Larson and Williams, 1987, op. cit.

62. Burnham, J.B., Giuliani, M.H., and Moeller, D.J., *Development, Installation, and Operating Results of a Steam Injection System (STIG) in a General Electric LM 5000 Gas Generator*, Paper 86-GT- 231, presented at the International Gas Turbine Conference and Exhibition, Dusseldorf, West Germany, American Society of Mechanical Engineers, June 8-12, 1986.

63. Oganowski, G., *LM 5000 and LM 2500 Steam Injection Gas Turbine*, presented at the International Gas Turbine Congress, Tokyo, Gas Turbine Society of Japan, Oct. 26-31, 1987.

64. Ibid.

65. Soroka, G.E. (Bechtel Eastern Power Corp.), *Modular Remotely Operated Fully Steam-Injected Plant for Utility Application*, paper presented at the ASME Cogen-Turbo Conference, Montreux, Switzerland, American Society of Mechanical Engineers, Sept. 1987.

66. Brandt, op. cit.

67. The intercooled STIG could also be used for cogeneration. An ISTIG unit based on the LM-5000 would produce about twice as much steam in the heat recovery steam generator as a STIG unit, largely because the turbine exhaust flow would be increased 26%, and its temperature would be increased from 395 to 441°C. [See Eng. Dept., Pacific Gas and Electr. Co. 1984. Scoping study: LM5000 steam-injected gas turbine. Based on work performed by the Mar. and Ind. Engine Proj. Dept. (Evendale, Ohio) of the Gen. Electr. Co.] About 44% of the produced steam would have to be injected for NO$_X$ control and cooling of the combustor. The rest, some 27,200 kg/hour at 41.3 bar plus 15,400 kg/hour

at 13.8 bar, could be used for process. In the full cogeneration mode the electrical output of the ISTIG unit would be 97 MW, and electricity production would represent 42.3% of the higher heating value of the fuel input (personal communication from M. Horner, General Electric Company, September 1988).

68. The LM-8000 is based on an upgraded version of the CF6 jet engine, the CF6-80C2, for which production totalled 110 engines in 1987 and is expected to be 260 engines in 1988. (See Ungeheuer, F., "They Make Good Things for Flying," *Time*, May 2, 1988, p. 55.)

69. Horner, M.W. [Mar. and Ind. Eng. and Serv. Div. (Cincinnati, Ohio) of the Gen. Electr. Co.]. 1988. Position statement - intercooled steam-injected gas turbine. Testimony presented at the Committee Hearing for the 1988 Electricity Report of the Calif. Energy Comm., held at the So. Calif. Edison Co., November 21-22, 1988.

70. Rice, I.G., "Combined Cycle, Reheat Gas Turbine, and Steam Cooling," testimony before the California Energy Commission, Hearing Docket No. 89-FR-1, Sacramento, Calif., July 17, 1989.

71. Janes, C.W., "Increasing Gas Turbine Efficiency Through the Use of a Waste Heat Methanol Reactor," *Proc. 14th Intersoc. Energy Convers. Eng. Conf.* 2: 1968-1972, 1979.

72. Tsuruno, S., and Fujimoto, S., "Gas Turbine Cycle with Steam Reforming of Methanol," Paper 87-Tokyo-IGTC-83, presented at the 1987 Tokyo International Gas Turbine Congress, Tokyo, Japan, Gas Turbine Society of Japan, Oct. 26-31, 1987.

73. Klaeyle, S., Laurent, R., Nandjee, F. May 31 - June 4, 1987. *New Cycles for Methanol-Fueled Gas Turbines*. Paper 87-GT-12. Presented at the Int. Gas Turbine Conf. and Exhib. Anneheim, Calif.: Am. Soc. Mech. Eng.

74. Mar. and Ind. Eng. and Serv. Div. (Cincinnati, Ohio) of the Gen. Electr. Co., *Advanced Chemically Recuperated Gas Turbine Cycle Evaluation Project: Preliminary Assessment of the System Concept*, Work sponsored by the Calif. Energy Commission, December 1988.

75. Rice, op. cit.

76. Engineering Department, Pacific Gas and Electric Co., *Scoping Study: LM5000 Steam-Injected Gas Turbine*, based on work performed by the Mar. and Ind. Engine Proj. Dept. (Evendale, Ohio) of General Electric Co., 1984.

77. Cohn, A., *Steam-Injected Gas Turbines Versus Combined Cycles*. *EPRI J.* Vol., 13, No. 5, 1988, p. 40.

78. See note 41.

79. Cohn, op. cit.

80. For an ISTIG unit operated on natural gas about two-thirds of the water vapor in the exhaust stream would have to be recovered to reduce makeup water requirements to zero, at zero relative humidity. The rest of the water vapor in the exhaust is a product of fuel combustion.

81. Corman, J.C., *System Analysis of Simplified IGCC Plants,* report prepared for the US Department of Energy by General Electric Corp. Research and Development, Schenectady, N.Y., 1986.

82. Argonne National Laboratory, *An Assessment of the Natural Gas Resource Base of the United States.* Washington, D.C.: Off. Policy, Planning, and Analysis, US Department of Energy, 1988.

83. Adelman, M.A., and Lynch, M.C., "Natural Gas Supply in Western Europe." *Western Europe Natural Gas Trade, Final Report.* Cambridge, Mass., International Natural Gas Trade Project, Center for Energy Policy Research, Energy Laboratory, Massachusetts Institute of Technology, 1986.

84. Williams, R.H., and Larson, E.D., "Expanding Roles for Gas Turbines in Power Generation," *Electricity: Efficient End-Use and New Generation Technologies, and Their Planning Implications,* eds. T.B. Johansson, B. Bodlund, and R.H. Williams, Lund, Sweden, Lund University Press, 1989.

85. Peters, W., "Coal Gasification Technologies for Combined Cycle Power Generation." *Electricity: Efficient End-Use and New Generation Technologies, and Their Planning Implications,* eds. T.B. Johansson, B. Bodlund, and R.H. Williams, Lund, Sweden, Lund University Press, 1989.

86. Williams and Larson, 1989, op. cit.

87. The technology proven at Cool Water achieves sulfur removal through the use of a scrubber. Though effective in removing sulfur, this approach requires a cooling of the gas exiting the gasifier and thus a loss of energy efficiency. More advanced techniques that remove the sulfur without cooling the gas ("hot-gas cleanup") are in an advanced state of development but are unproven at commercial scales. (See R.H. Williams and E.D. Larson, 1989, op. cit.)

88. R.H. Williams and E.D. Larson, 1989, op. cit.

89. Fluor Engineering, Inc., *Planning Data Book for Gasification-Combined Cycle Plants: Phased Capacity Additions,* EPRI AP-4395. Palo Alto, Calif., Electric Power Research Institute, 1986.

90. Catina, J.L., and Fortune, H.J. Jr. (Virginia Power), Soroka, G.E. (Bechtel Eastern Power Corp.), *Repowering Chesterfield 1 and 2 with Combined Cycle,* Paper 87-GT-12, presented at the International Gas Turbine Conference and Exhibition, Anaheim, Calif., American Society of Mechanical Engineers, May 31-June 4, 1987.

167

91. Larson, E.D., Svenningsson, P., and Bjerle, I., "Biomass Gasification for Gas Turbine Power Generation," *Electricity: Efficient End-Use and New Generation Technologies, and Their Planning Implications*, eds. T.B. Johansson, B. Bodlund, and R.H. Williams, Lund, Sweden, Lund University Press, 1989.

92. Williams, R.H., *Biomass Gasifier/Gas Turbine Power and the Greenhouse Warming*, paper presented at the Expert Seminar on Energy Technologies for Reducing Emissions of Greenhouse Gases. Paris, France: IEA/OECD, April 1989.

93. Hubbard, H.M., "Photovoltaics Today and Tomorrow," *Science* Vol. 244, 1989, pp. 297-304.

94. Carlson, D.E., "Low-Cost Power from Thin-Film Photovoltaics," *Electricity: Efficient End-Use and New Generation Technologies, and Their Planning Implications*, eds. T.B. Johansson, B. Bodlund, and R.H. Williams, Lund, Sweden, Lund University Press, 1989.

95. For utility gas turbines consuming more than 100 million Btu per hour (approximately 10 MW) the standard is 75 ppm x (n/25) by volume (@ 15% O_2), where n is the turbine efficiency in percent. The standard for all other continuous-duty gas turbines consuming more than 10 million Btu per hour and producing less than 30 MW is 150 ppm x (n/25).

96. The toughest proposed standards in the US are those proposed in August, 1988, by the South Coast Air Quality Management District (SCAQMD) in California: an emission level of 9 ppm x (n/25) for all new stationary gas turbines with capacities greater than 0.3 MW (tentative adoption date: January 1989) and an emission level of 12 ppm x (n/25) for existing turbines, within 30 months of the adoption date. The proposed standards are part of the proposed SCAQMD plan to bring the greater Los Angeles area into compliance with the US ambient air quality regulations established under the Clean Air Act. (See South Coast Air Quality Management District (El Monte, Calif.). May 13, 1988. Proposed Rule 1134 -- *Control of Oxides of Nitrogen Emissions from Stationary Sources*.)

97. Fraize and Kinney, op. cit.

98. Rice, op. cit.

99. Sidebotham, G.S., and Williams, R.H., *Preliminary Report on NO_x and Cogeneration in New Jersey*, Princeton, N.J., Center for Energy and Environmental Studies, Princeton University, 1989.

100. Office of Technology Assessment, *Acid Rain and Transported Air Pollutants: Implications for Public Policy*. Washington, D.C., 1984.

101. Williams and Larson, op. cit.

102. Ibid.

103. World Bank, *The Energy Transition in Developing Countries*, Washington, D.C., 1983.

104. Energy Information Administration, 1988, op. cit.

105. Ibid.

106. The breakeven price is determined by setting the levelized busbar cost from a natural gas-fired ISTIG unit equal to the operating cost of an existing steam-electric plant, assuming a 46% capacity factor for the ISTIG unit, the same fuel price for both plants, and an O&M cost of 4.0 mills per kWh for the existing steam plants [the average value in the US in 1985 (Energy Inf. Admin., 1987, op. cit.)].

107. Energy Information Administration, 1988, op. cit.

108. California Energy Commission Staff, *Resource Case Analysis Report*, testimony presented at the Committee Hearing for the 1988 Electricity Report of the California Energy Commission, held at Southern California Edison Co., November 21-22, 1988.

109. Schramm, G., "The Changing World of Natural Gas Utilization," *Natural Resources Journal*, Vol. 24, 1984, pp. 405-36.

110. See note 34.

111. See note 36.

ACKNOWLEDGMENT

The authors gratefully acknowledge partial support from the Office of Energy of the US Agency for International Development for the research upon which this chapter is based.

6

The Use of Natural Gas in the Nitrogen Fertilizer Industry

William F. Sheldrick

Introduction

More than 70% of world production of ammonia depends on natural gas, and this percentage is expected to increase through the year 2000 and beyond. Ammonia is the building block from which more than 95 of the world's nitrogen products are derived, and nearly 90% of the world's ammonia production is dedicated to the production of nitrogen fertilizers. Consequently the production of food in the world is highly dependent on the availability and cost of natural gas.

Natural gas is also used to some extent in the manufacture of phosphate and potash fertilizers but mainly as a source of fuel for the generation of steam and electricity. In the case of nitrogen fertilizers, natural gas, apart from providing fuel energy, also provides feedstock for the production of hydrogen used for the synthesis of ammonia. About one third of total energy needed is as fuel and the remaining two thirds as feedstock. A natural gas-based nitrogen fertilizer complex requires approximately 50-60 thousand cubic feet of natural gas (about 50-60 million Btu's) to produce one metric ton of nitrogen nutrient, either as solid urea or ammonium nitrate. To produce one nutrient ton of potash fertilizer or one nutrient ton of phosphate fertilizer requires only about one tenth of the equivalent energy to produce one ton of nitrogen nutrient as urea.

Without a doubt the development of the large-scale natural gas-based ammonia plant has played an important part in helping to meet our food needs over the last two decades and particularly in ensuring the success of the "green revolution." The availability of inexpensive natural gas is the basis of domestic nitrogen fertilizer industries that helped them increase their agricultural outputs. The wide availability of natural gas in many parts of the world at prices well below the price of other forms of energy, and the savings in both energy and investment costs when using natural gas compared with other alternative feedstock sources have resulted in a steady reduction in nitrogen fertilizer prices over the past two decades. The price of nitrogen fertilizers in constant dollar terms has been reduced to 35% of its value in 1965, whereas in the case of phosphate and potash fertilizers the corresponding figure is 70%.

Natural gas will continue to play an increasingly important role in the nitrogen fertilizer industry in the future. The need to increase food production to meet growing world population, which in absolute terms is growing at its highest rate ever, means that nitrogen fertilizer use must be increased at least 2-3% per year. Many of the benefits of the high yield varieties have now been fully realized; water availability, soil erosion and environmental problems will increasingly constrain food production so we shall become more dependent on fertilizer use to meet our agricultural needs. Although the biological fixation of atmospheric nitrogen in theory offers a cheap and promising source of nitrogen for agriculture, it is unlikely that further developments in this area will have any significant impact on the increasing need for nitrogen fertilizers in the next decade or so. Therefore, natural gas will continue to be the main ingredient for the production of nitrogen fertilizers. The remainder of this chapter reviews the nitrogen fertilizer industry with particular reference to the use of natural gas.

Review of the Nitrogen Fertilizer Industry

Historical Aspects

At the beginning of the twentieth century, a growing population placed increasing emphasis on the need to augment agricultural production by increasing nitrogen fertilizer use, and considerable research was initiated to develop new processes for the fixation of atmospheric nitrogen into forms that could be used for the manufacture of nitrogen fertilizers. As a result, three new processes to fix nitrogen were developed and operated commercially. In 1903 the Arc process was commissioned in Norway; in it nitrogen and oxygen were combined to form a nitric oxide arc. At a lower temperature the nitric oxide was reacted to form nitrogen dioxide, which was then reacted with water, and more nitrate was produced. Another process was developed at the same time to produce calcium cyanamide. Calcium carbide produced by the reaction of lime with coke in an electric furnace was converted to calcium cyanamide by reacting it with nitrogen extracted from the air.

The most important development in nitrogen fixation, however, was the Haber Bosch process that was introduced on a commercial scale in Germany in 1913. This process was based on the catalytic reaction of hydrogen and nitrogen at high temperature and pressure; this basic process is still used today.

172

Until about 1960, most of the world ammonia manufacturing capacity was located in developed countries. Plants were small, used a wide variety of feedstocks and served local markets. There were very few ammonia plants in developing countries and only a small amount of fertilizer was imported by these countries. During the 1960s several major changes took place in the industry. Although the price of naphtha and fuel oil, two major feedstocks for ammonia production, rose sharply, generally cheap natural gas became much more available, particularly for new plants in developing countries. At the same time, advances in technology resulted in larger scale plants that could use centrifugal compressors, which in turn provided benefits of scale and energy savings.

The nitrogen industry expanded rapidly in the 1960s and 1970s as the demand for both fertilizer and industrial nitrogen increased. The structure of the industry has changed as both the production and consumption of nitrogen fertilizers grew in all regions. By 1988 the major producers and consumers of nitrogen fertilizers were USSR, China, US and India. About 12% of ammonia production went to industrial use mainly in the developed market economies.

Feedstock for Ammonia Production

Before World War II coal and coke accounted for more than 90% of the 3 million tons per year of nitrogen produced as ammonia at that time. By 1960, with a world capacity of ammonia of about 10 million tons per year, natural gas had overtaken coal and coke as the main feedstock dye mainly because of the availability of cheap gas in the US. Many other developed countries' recent new capacity was based on fuel oil, but by this time the most preferred feedstock for the industrial countries that were not gas-based was naphtha, as it represented the cheapest available fraction of the oil barrel. The development of the ICI process for the steam reforming of naphtha resulted in many new plants in all parts of the world, not only for ammonia, but also for methanol and town gas. However, the resulting increase in the demand for naphtha and the oil crisis of 1973 increased the relative price of naphtha compared with other feedstocks, making it less competitive; many plants were idled or closed permanently. Since about 1975 almost all new plants have been based on natural gas, apart from the special case of China where coal still continues to be a major feedstock. About 73% of world ammonia capacity is based on natural gas, but if as shown in Figure 6.1 the Chinese small plants are excluded from the total the percentage of gas-based plants would increase to more than 78%.

Total energy needs and investment costs when using natural gas for ammonia production are usually lower than for other feedstocks. Based on the present projections for reserves and opportunity costs for gas,

173

Figure 6.1. Feedstock Source for Ammonia Capacity Worldwide

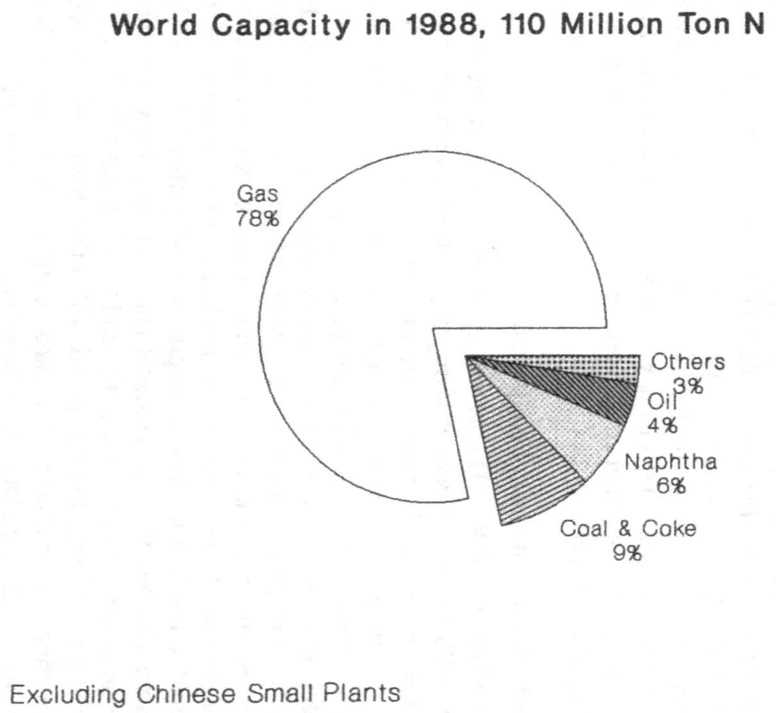

World Capacity in 1988, 110 Million Ton N

Gas
78%

Others
3%
Oil
4%
Naphtha
6%
Coal & Coke
9%

Excluding Chinese Small Plants

particularly in developing countries, it seems likely that natural gas will continue to increase its position as the preferred feedstock for ammonia production.

Ammonia Production Capacity

Information on the development of nitrogen production in Figure 6.2 shows how the industry has grown since the beginning of the century, particularly during the last two decades. Up to about 1950 the industry had developed relatively slowly, reaching about 4.5 million tons of nitrogen production per year. In the next ten years nitrogen production increased to just over 10 million tons, and by 1960 and 1989 many developing countries recognized the need to establish nitrogen fertilizer capacity to protect their agricultural programs, particularly after massive hikes in fertilizer prices that took place in 1973. Figure 6.3 shows how investment in ammonia capacity has taken place since 1965.

World Nitrogen Consumption

World nitrogen consumption has increased about ten-fold over the last three decades. There have been two major recessions in demand. The first recession took place in 1975 after the major price hike during the so-called Fertilizer Crisis in 1974. The second occurred in the early 1980s as a result of the global recession and the Payment in Kind (PIK) program in the US. World nitrogen consumption since 1960 is given in Table 6.1. Figure 6.4 illustrates the historical growth of fertilizer demand by region.

Table 6.1. World Nitrogen Consumption 1960-1990 (Million Tons N)

	59/60	64/65	69/70	74/75	79/80	84/85	89/90
Fertilizers	9.2	15.3	28.7	38.6	57.3	70.5	80.3
Industrial	3.9	6.0	7.0	8.1	9.2	9.6	10.1
Total	13.1	21.3	35.7	46.7	66.5	80.1	91.4
Average % Increase Over		10.2	10.9	5.5	7.3	4.0	2.5[a]

[a]Forecast.

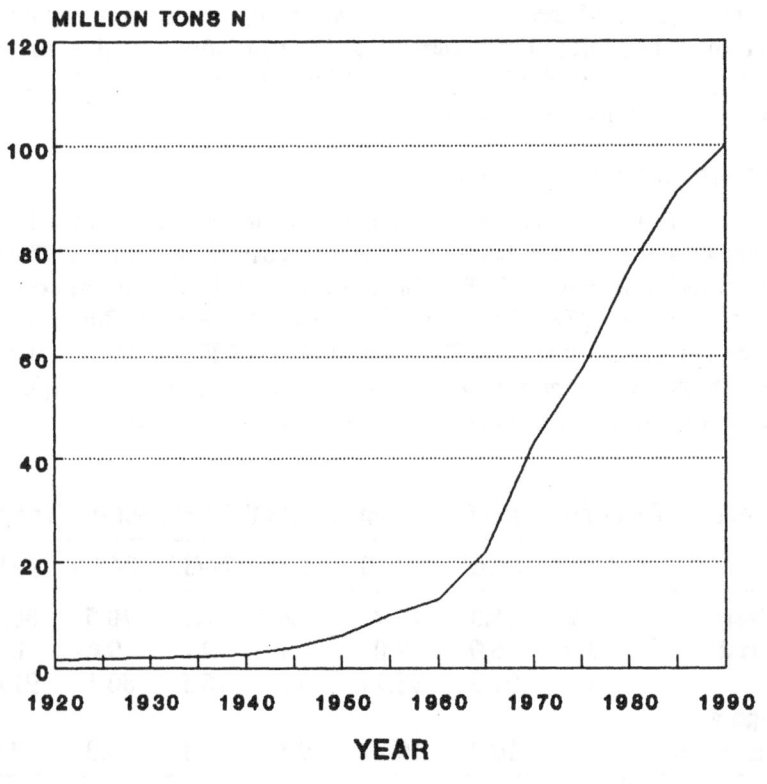

Figure 6.2. World Nitrogen Production

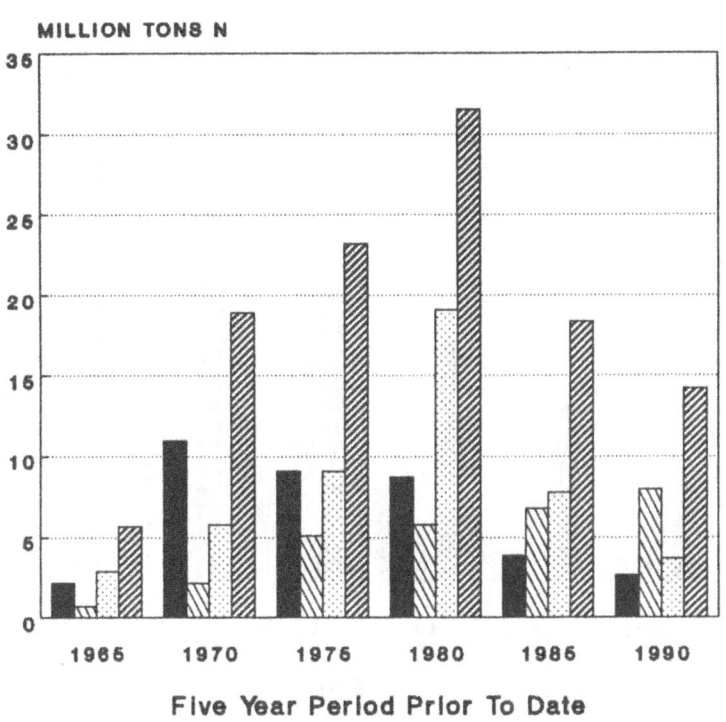

Figure 6.3. **Ammonia Plant Construction for Five-Year Periods (1960-1990)**

MILLION TONS N

Five Year Period Prior To Date

Developed M.E. Developing M.E. Cent. Plan.E. World

Including Chinese Small Plants

177

Figure 6.4.   World Nitrogen Fertilizer Demand

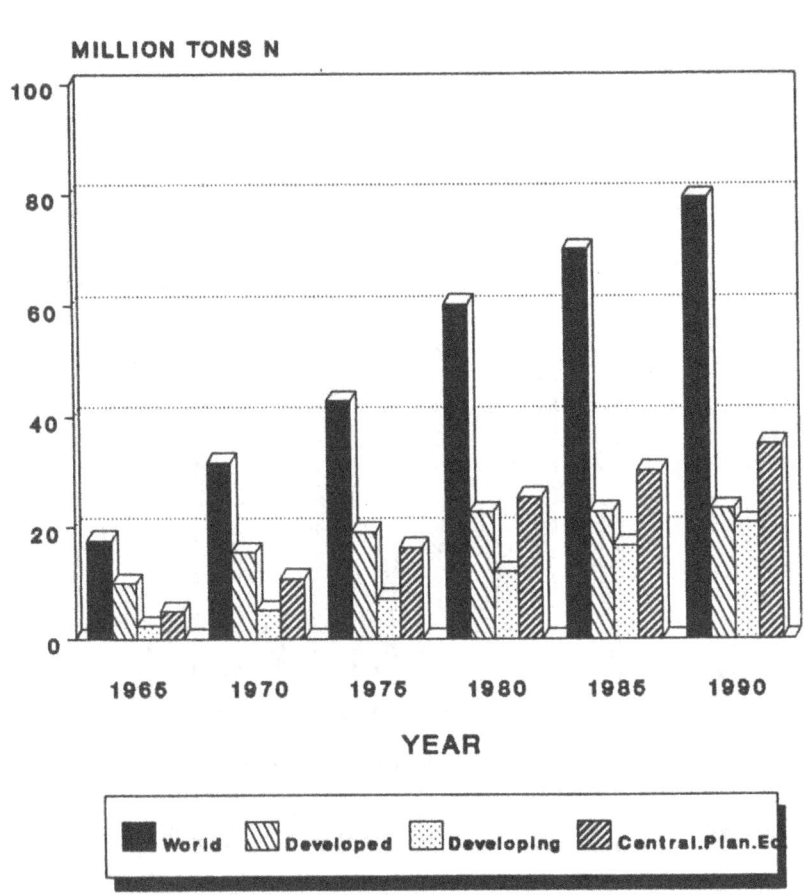

Between 1960 and 1970 nitrogen fertilizer growth averaged 10% per year but has been falling steadily, and between 1984/85 and 1989/90 it is expected to average only 2.5%. A breakdown of nitrogen fertilizer consumption for the major economic regions is given in Figure 6.4.

Industrial nitrogen demand in 1987/88 totaled about 9.9 million tons, which was 12% of total nitrogen consumption of 85.9 million tons. About 70% of industrial demand for ammonia occurs in the developed market economies, 25% in the centrally planned economies and 5% in the developing market economies. The four major consumers are the US, USSR, Japan and Western Europe. In Japan about 60% of all nitrogen products consumed go to industrial use.

World Nitrogen Fertilizer Trade

General. During the last twenty-five years there has been a major increase in both production and consumption of nitrogen fertilizers in almost all regions of the world. The fact that these have not always increased at the same rate has resulted in a steady increase in international trade in nitrogen fertilizers. In 1959/60 exports of nitrogen fertilizers totaled about 2.6 million tons and by 1986/87 had increased to 17.4 million tons -- about 23% of total world production. The regional percentage share of nitrogen fertilizer trade shown in Table 6.2 indicates its wide distribution. Many large countries or regions are both large exporters and importers of fertilizers, particularly Western Europe and North America. During the period 1959/60 to 1987/88 exports of nitrogen fertilizers outside the developed market economies increased from virtually zero to more than 40%. The major countries involved in fertilizer trade are given in Table 6.3.

Nitrogen Fertilizer Products Traded. The types of nitrogen fertilizer products traded have changed a great deal in the last two decades. The most important aspect of this change has been the increase in urea to become the most important fertilizer traded. Urea is a relatively easy and convenient material to produce as it can be integrated with ammonia production, and producers for the export market with cheap natural gas have usually invested in urea production facilities. Because of its favorable economics of production and high concentration, urea seems likely to increase its market share even further. The major exporters of urea are USSR, Romania, the Netherlands, and Indonesia. Although ammonium sulphate was once a very popular fertilizer, its use has declined due to high cost of production and low analysis. It is now recovered mainly as a by-product of the synthetic textile industry, and its main source is Western Europe and Japan.

179

Table 6.2. Share Distribution of Nitrogen Fertilizer Trade (%)

	Imports		Exports	
	1985/86	1986/87	1985/86	1986/87
Developed Market Economies	49.7	52.8	52.5	52.2
North America	23.1	22.5	22.1	23.0
Western Europe	24.9	28.1	28.1	27.5
Oceania	0.9	0.9	0.3	0.2
Others	0.8	1.3	2.0	1.5
Developing Market Economies	33.2	31.6	17.8	19.8
Africa	3.5	3.4	0.7	0.8
Latin America	6.6	8.5	2.2	2.7
Near East	7.3	7.0	9.6	9.7
Far East	15.8	12.7	5.3	6.6
Centrally Planned Economies	17.1	15.6	29.7	28.0
Socialist Asia	14.2	12.5	0.1	0.3
Eastern Europe incl. USSR	2.9	3.1	29.6	27.7
Total	100.0	100.0	100.0	100.0
Trade Million Tons Nitrogen	16.53	16.80	15.37	17.44

Source: *FAO Fertilizer Handbook 1987*.

Ammonium nitrate is produced mainly in Western Europe where it is the principal source of nitrogen fertilizers. Most trade in ammonium nitrate takes place as interregional trade in Western Europe, but some is exported to other regions.

A good part of the production of calcium ammonium nitrate is concomitant with nitrophosphate fertilizer production, mainly in Western Europe, but because of its relatively low concentration it does not form a major part of long-distance trade. A significant quantity of nitrogen is also shipped as diammonium phosphate.

Before 1960 there was virtually no trade in ammonia, but by 1987 world ammonia trade had grown to 8.5 million tons. In the late 1970s there was a rapid increase in trade, mainly to serve the increasing needs of Europe. Transatlantic trade to bring the cheap ammonia from the new large plants in the US was also accompanied by considerable interregional trade in Europe. The structure of the industry has changed, however, and the US is now a net importer of ammonia, although a good part of this ammonia is

Table 6.3.   Country Trade in Nitrogen Fertilizers (000 Tons N)

Country	Imports	
	1985/86	1986/87
US	3627	3543
China	2074	1757
France	970	1244
India	1616	1106
Germany FR	795	914
Iran	414	457
UK	431	452
Belgium	358	378
Thailand	287	365
Turkey	300	452

Country	Exports	
	1985/86	1986/87
US	1879	2460
USSR	2042	2409
Canada	1510	2460
Netherlands	1241	1507
Romania	1362	1160
Belgium	985	989
Indonesia	338	696
Italy	346	528
Germany FR	339	476
Germany DR	291	437

Source: *FAO Fertilizer Yearbook 1987.*

processed and exported as diammonium phosphate. In 1988 the US and Western Europe were the two main markets accounting for nearly 80% of total world ammonia trade. The USSR has become the major ammonia exporter with 25% of all ammonia exports in 1987.

Nitrogen Fertilizer Prices

Prices for ammonia and urea shown in Figure 6.5 indicate wide price fluctuations in the last twenty years. Following an increase of prices in the mid-1960s there was a period of over-supply when prices slumped. In 1974 there was a return to balance with a very sharp rise in prices that were affected by a major increase in energy prices and fears of a major

Figure 6.5. Ammonia and Urea Export Prices

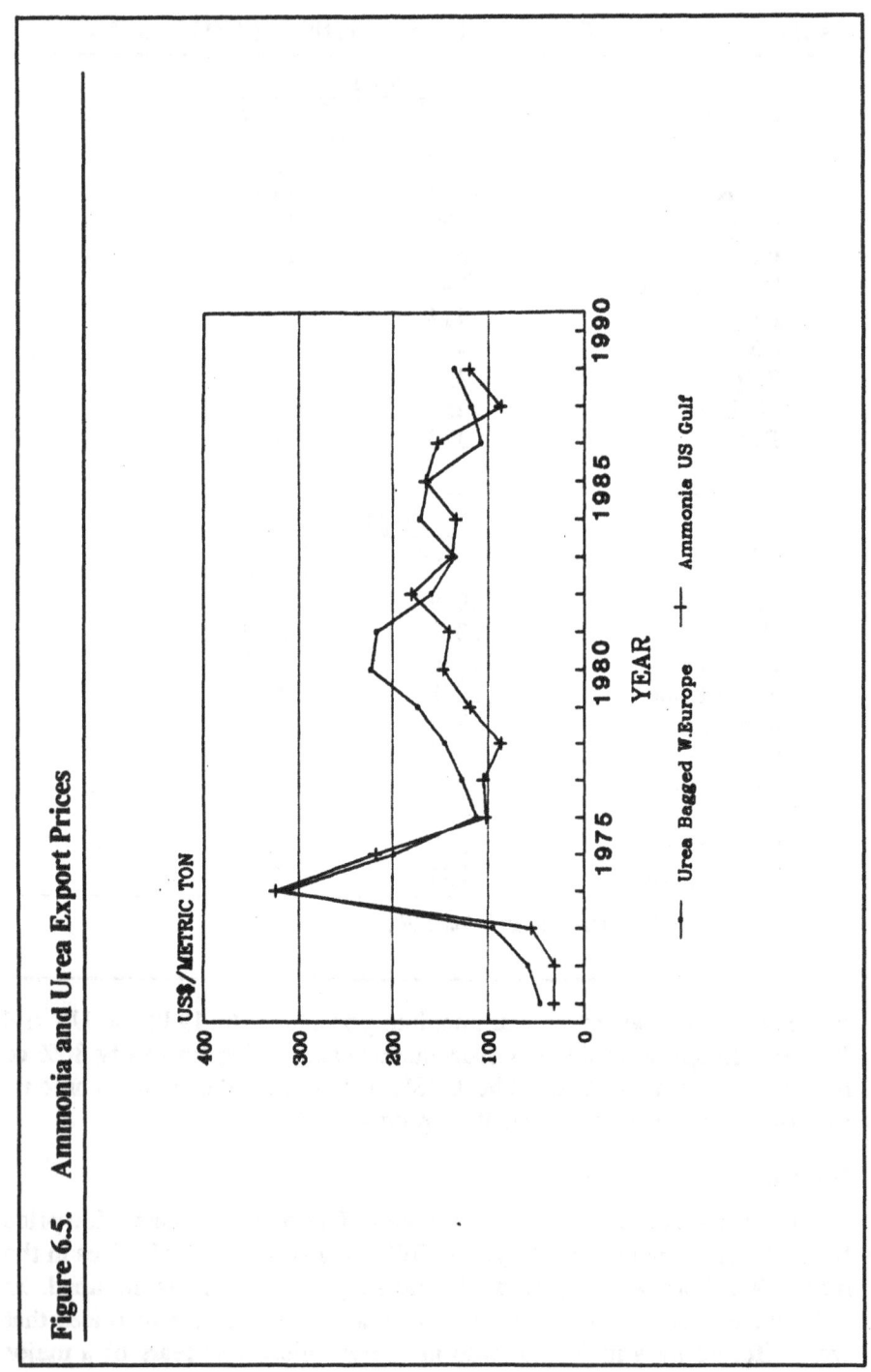

fertilizer shortage at that time. Prices at the end of 1974 rose to unprecedented heights as many developing countries fearful of a forthcoming shortage of fertilizers bought more than their immediate needs. This shortage, however, was short-lived because the rapid increases in prices slowed down. Prices fell in the following years, particularly as new capacity that had been constructed during the shortage came on stream.

In 1979 a surge in nitrogen demand brought about a more balanced situation and increased prices. This was particularly so for urea where capacity was more limited. During the mid-1970s there had been considerable investment in ammonia plants in the USSR, and when this capacity came on stream it helped constrain ammonia prices relative to urea prices. This balanced situation continued through 1981, but a depressed world agricultural situation stagnated fertilizer demand at a time when increasing flows of ammonia and urea were becoming available from Eastern Europe. Between 1981 and 1983 urea prices almost halved. In order to sell their increasing surplus of ammonia and urea to obtain sorely needed hard currency, Eastern Europe continued to displace other producers from the export market by reducing export prices below the cash costs of many US fertilizer companies, and between 1982 and 1983 more than 5 million tons of ammonia capacity, mainly in the US, were idled.

In 1983 it appeared that the bottom of the price cycle had been reached, and prices started to rise again in the face of strong demand, that increased nitrogen consumption by 10% in 1983/84. Demand increased again in 1984/85 by 5%, and prices continued to improve. After mid-1985 ammonia and urea prices fell as new supply potential became available. The depressed world agricultural situation, the impact of the US 1985 Farm Bill and the fact that some part of sales in the previous two years had gone into stocks were the main reasons for check in demand. The urea market in particular was dominated by two major purchasers: China and India. The failure of China to purchase major quantities of urea in 1985 and 1986 after its large purchases in 1984 depressed the market. In 1986 and 1987 prices of ammonia and urea remained low, but in 1988 they started to increase due to a gradual tightening of the supply situation and also because of a significant increase in consumption in North America. In 1988/89 farmers in the US endeavored to make up for the 1988 drought by increased plantings in the following season, but unfavorable spring planting conditions restricted fertilizer use and prices again declined.

Profiles of Major Country and Regional Ammonia Producers

A list of major ammonia producers is given in Table 6.4, and a short review of the nitrogen industries in these countries is also provided.

USSR. After many years as the world's leading ammonia producer the US was overtaken by the USSR in 1978, and it is projected that by 1993/94 the ammonia capacity in the USSR will be more than twice that in the US. With a large natural gas resource base, the USSR has more than doubled its capacity in the last decade, whereas capacity in the US has declined slightly. During the 1970s the Soviet Union ordered a total of 40 large ammonia plants mainly from the US, Western Europe and Japan. By the mid-1980s these plants had all come on stream, making the USSR the world's largest producer and exporter of nitrogen fertilizers and intermediates. More than 90% of ammonia production is now based on natural gas. In 1987 the country exported about 2.0 million tons of urea and about 2.2 million tons of ammonia.

The Soviet ammonia export commitments grew initially as a result of barter and buy-back deals with Western Europe and the US. Facilities were constructed for exporting ammonia at Ventspils in the Baltic and at Odessa in the Black Sea. The main facility of Odessa is supplied from ammonia complexes at Toglyatti and Gorlovka by pipeline. The USSR, with gas deposits exceeding 36,000 billion cubic meters, has the largest resources in the world. About 75% of Russian-explored line reserves are located in West Siberia, and a major pipeline has been built to bring this gas to the West. Other gas developments are taking place in Turkmenia and Astrakhan. The USSR has considerable potential to increase significantly its ammonia production based on gas, and with capacity expansions planned, it is expected to remain the world's leading nitrogen products exporter through the mid-1990s at least.

China. In 1949 China had only two small plants producing about 6,000 tons of N as ammonium sulphate, but in 1988 there were about 1400 ammonia plants with down-stream nitrogen fertilizer facilities. The development of China's nitrogen fertilizer industry began in the 1950s to produce ammonia almost exclusively from locally produced anthracite.

Table 6.4. Major Country Ammonia Capacities
(Million Tons N per Year)

Country	1973/74	1980/81	1987/88	1993/94[*]
USSR	10.2	19.2	24.9	26.0
China	6.3	16.0	17.9	20.8
US	12.8	15.7	12.8	12.8
India	2.2	4.9	7.7	10.0

[*]Projected.

Figure 6.6.   Feedstock Sources for Ammonia Capacity (China)

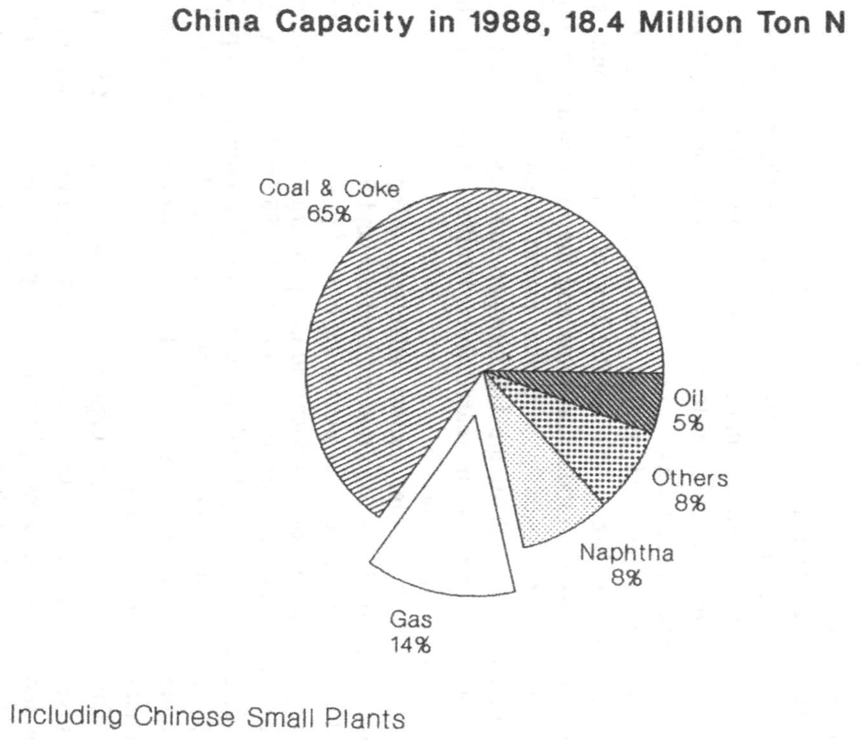

China Capacity in 1988, 18.4 Million Ton N

Coal & Coke
65%

Oil
5%

Others
8%

Naphtha
8%

Gas
14%

Including Chinese Small Plants

There are now more than 1300 of these small plants each producing 5000 to 15000 tons N per year as ammonia. The main products are ammonia liquor and ammonium bicarbonate, although some of these plants also produce urea and ammonium nitrate. In the 1970s China purchased thirteen 1000 tpd ammonia plants to use a variety of feedstocks. In addition about 50 medium-size plants were designed and built using domestic know-how. Even though ammonia and urea production has increased rapidly it has not been sufficient to keep up with consumption, and China is likely to remain a major importer of nitrogen fertilizers well into the 1990s.

Ammonia production in China is now based on a wide range of feedstocks, and the mix in 1988 is shown in Figure 6.6. For future production of ammonia China has abundant resources of coal, oil and natural gas, and depending on location China will continue to use natural gas for ammonia production.

US. Until about 1980 the US was the largest producer of ammonia and nitrogen fertilizers. The availability of low-cost natural gas, low investment costs for production units and the introduction of high yield varieties of corn together with the export boom of the 1970s had stimulated the growth of the US nitrogen industry. Between 1960 and 1980 ammonia capacity increased from 4 million tons (N) to almost 16 million tons (N) per year.

After 1980, several factors adversely affected the US ammonia industry. A world recession that reduced demand for agricultural products and hence fertilizers and an increasing cost of producing ammonia due to rising gas prices made it difficult to compete with imports. The very low prices and intense competition from the USSR and Eastern European producers caused nearly 5 million tons (N) of US ammonia capacity to be idled in the early 1980s, and more than 3 million tons will remain permanently idled. Within this period the US changed from a net exporter to a net importer of nitrogen.

The US had been a net importer of ammonia (distinct from nitrogen) for several years previously, but most of this ammonia was processed and exported as diammonium phosphate. Imported ammonia comes mainly from Canada, the Caribbean countries and Eastern Europe. It seems likely that the US will meet increasing demands of ammonia by importing rather than building new capacity, although some existing plants will be refurbished to reduce energy needs and at the same time expand existing capacity. In 1988 there were forty producers of ammonia with a nameplate capacity of about 13.0 million tons.

India. India has made tremendous progress in developing nitrogen fertilizer production. With nearly forty plants in operation, India is ranked

Figure 6.7.   Feedstock Sources for Ammonia Capacity (India)

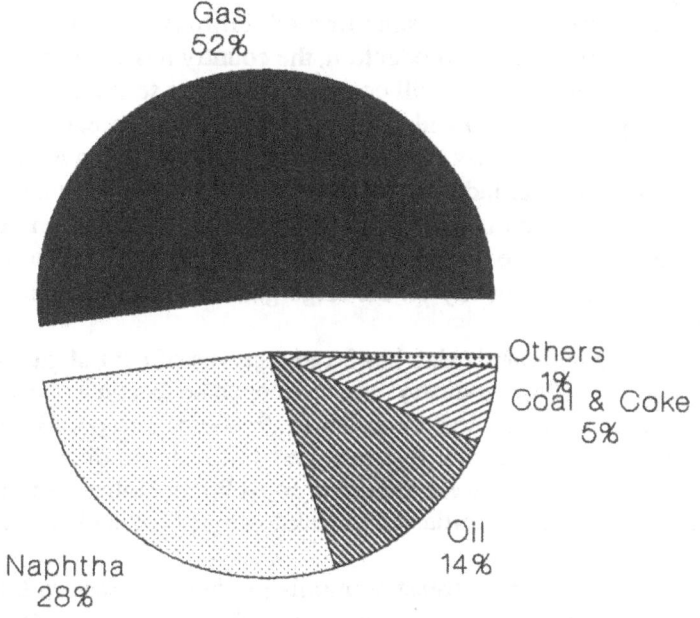

India Capacity in 1988, 8.6 Million Ton N

Gas
52%

Others
1%

Coal & Coke
5%

Oil
14%

Naphtha
28%

as the world's fourth largest producer. In 1988 Indian ammonia capacity was 8 million tons, and, with several more plants planned to come on stream, this capacity should reach 10 million tons by 1993. A wide range of feedstocks is used in India as indicated in Figure 6.7.

Most of the older capacity is based on naphtha (46% of total ammonia capacity in 1983). Following the development of the Bombay High associated gas field most new plants have been based on gas, so that by 1988 the percentage of ammonia capacity based on gas had increased to 52%. Of eleven new plants either recently commissioned, under construction or planned, ten are based on natural gas from the Bombay High field.

Because India's consumption of fertilizers has steadily outpaced its supply from domestic production, the country has been a major importer of ammonia and urea and will continue to remain so until the increase of new capacity can catch up with demand, sometime in the early 1990s.

Other Countries. Other countries with a major involvement in the nitrogen fertilizer industry include Canada, the Netherlands, Indonesia and Romania. Canada is a major producer and exporter of ammonia and urea. There are twelve companies producing about 3 million tons (N) as ammonia based mainly on the large gas reserves in British Columbia and Alberta.

The Netherlands also has large reserves of natural gas, and the Dutch nitrogen fertilizer industry has developed to become the major nitrogen fertilizer producer and exporter in Western Europe. Holland produced about 1.7 million tons (N) nitrogen fertilizer products in 1987/88, and about 80% of these were exported. In addition, Holland exported about 0.8 million tons of ammonia. The breakdown of feedstock source is presented in Figure 6.8.

Romania is the largest ammonia producer in Eastern Europe after the USSR and in 1988 had a capacity of 2.6 million tons (N). As a result, Romania is now one of the world's largest exporters of nitrogen fertilizers. Although its ammonia production capacity was originally based on domestic natural gas, Romania now imports natural gas from the USSR to meet its total needs.

Indonesia has also developed a large nitrogen fertilizer industry based on natural gas, and most of the production goes to the domestic market.

Ammonia Production from Natural Gas

General

The synthesis of ammonia is achieved most directly by fixing nitrogen with hydrogen at high pressure and temperature using a catalyst. The

Figure 6.8.   Feedstock Sources for Ammonia Capacity (W. Europe)

W. Europe Capacity in 1988, 14.6 Million Ton N

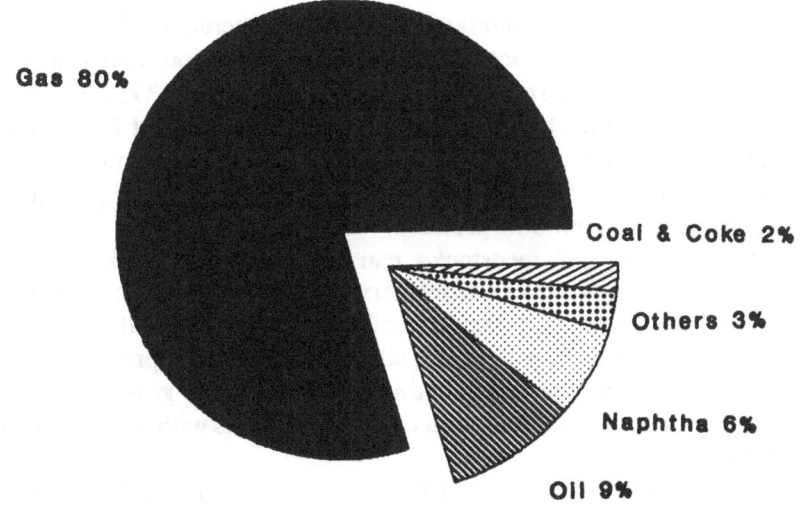

Gas 80%

Coal & Coke 2%

Others 3%

Naphtha 6%

Oil 9%

chemistry and thermodynamics of this reaction have been studied intensively and the steady improvement that has been achieved in the energy efficiency of modern ammonia processes is due mainly to the application of thermodynamics to the detailed plant design. The principles of the process for the production of ammonia have changed little in the last fifty years, and improvements in the process are not the result of any major technical breakthroughs but rather the integration of a number of energy conservation features.

Basically the ammonia process comprises three main steps:

- Synthesis gas preparation;

- Purification of the synthesis gas;

- Compression of the synthesis gas and ammonia synthesis.

The synthesis gas preparation involves the generation of hydrogen and the appropriate introduction of nitrogen. Purification consists of the removal of carbon monoxide (CO) and carbon dioxide (CO_2), the removal of catalyst poisons such as sulphur and the preparation of hydrogen and nitrogen in the stoichiometric ratio of $3H_2/N_2$. Ammonia synthesis covers the catalytic fixation of nitrogen at elevated temperature and pressure and the recovery of the ammonia product.

A wide range of feedstocks can be used for the production of hydrogen, such as electrolysis, refinery or coke oven gases, natural gas, light gasoline, crude and residual oil, and coal. Normally electrolytic hydrogen is expensive, so nearly all synthesis gas is produced by the gasification of liquid or solid hydrocarbons or from gaseous fuel. The process then requires the oxidation of these materials with oxygen or steam as follows:

$$C + 1/2O_2 \longrightarrow CO \qquad \text{(partial oxidation)}$$
$$C + H_2O \longrightarrow CO + H_2 \qquad \text{(water gas)}$$
$$CH(2n+2) + (n/2)O_2 \longrightarrow nCO + (n+1)H_2 \qquad \text{(partial oxidation)}$$
$$CH(2n+2) + (n/2)H_2) \longrightarrow nCO + (2n+1)H_2 \qquad \text{(reforming)}$$
$$CO + H_2O \longrightarrow CO_2 + H_2 \qquad \text{(shift conversion)}$$

Generally as the molecular weight of the feedstock increases, so does the complexity and cost of the process. The higher the H_2/C ratio of the fuel, the more economic the process. The partial oxidation process is used for the production of hydrogen from fuel oil, and the steam reforming process is used for processing naphtha and gaseous hydrocarbons such as methane.

There is a considerable amount of excellent literature describing in detail the production of ammonia using different feedstocks, thus the main purpose of the remainder of this section is only to highlight some of the major process features and changes that have recently taken place in the development of the ammonia process using natural gas in recent years.

Basically the production of ammonia comprises a succession of high temperature gaseous reactions, both exothermic and endothermic, into which heat is injected as fuel and from which waste heat must be recovered in a usable form (steam). By careful process and plant design that take into account the thermodynamics and kinetics of the reactions involved, the heat balance can be optimized. Progress during the last two decades, following the major developments of large plants using centrifugal compressors, has been largely along the lines of improving the design of the existing system, and to reduce energy consumption and capital costs. The progress that has been made in reducing energy needs can be seen in Figure 6.9. Energy consumption has been almost halved since 1963 when the first large single-stream ammonia plant using centrifugal machines to compress synthesis gas to elevated pressures was built by Kellogg.

The Technology of the
Natural Gas Steam Reforming Process

Although ammonia plants offered by major suppliers may vary considerably in detailed design and operating conditions, basically they all contain the following ten stages:

- Desulphurization;

- Primary reforming;

- Secondary reforming;

- Shift conversion;

- Carbon dioxide removal;

- Methanation;

- Ammonia synthesis.

Desulphurization. Natural gas is compressed to reformer pressure, and then the residual sulphur is removed to prevent catalyst poisoning later in the process. Sulphur is removed by adsorption on active carbon or more recently by treatment with zinc oxide.

191

Figure 6.9.   Energy Requirements for Ammonia Production

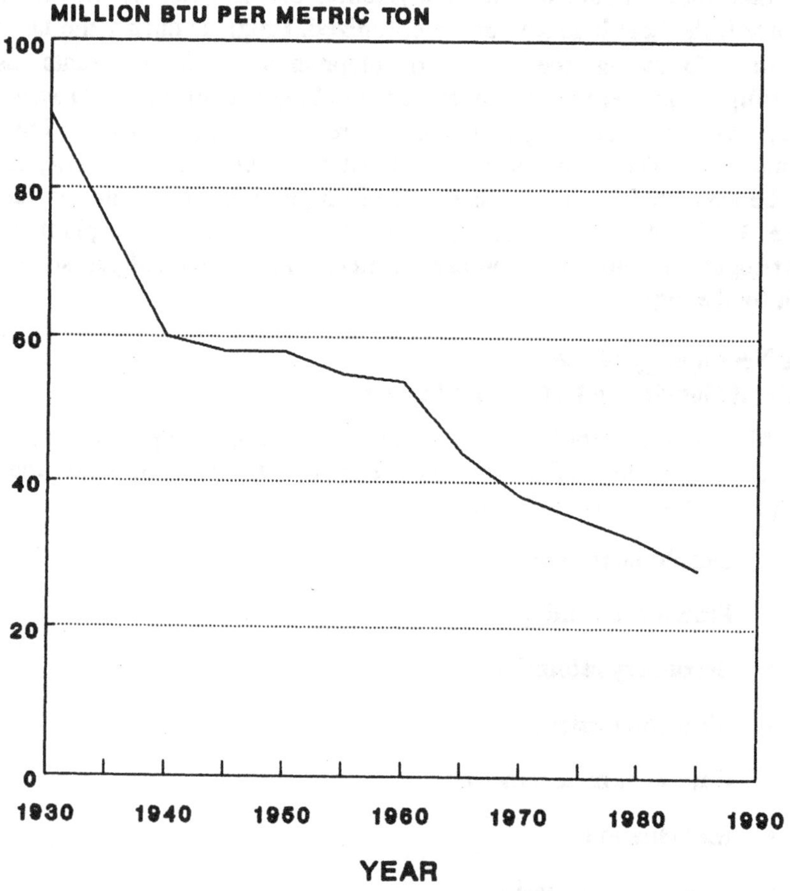

MILLION BTU PER METRIC TON

YEAR

Primary Reforming. In the primary reformer the bulk of the methane is converted into hydrogen and carbon monoxide by reaction with steam. The reaction is highly endothermic and energy is supplied to maintain the temperature at about 1000oC by burning fuel. The preheated feedstock and high pressure steam are mixed and passed through tubes containing a nickel-based catalyst in a reformer furnace at a pressure of 30-50 atm. The reformed gas leaving the tubes passes to the secondary reformer.

Secondary Reforming. The object of the secondary reformer is to complete the reaction of the unreacted methane from the primary reformer and to supply the required quantity of nitrogen (as air) for ammonia synthesis. Oxygen in the air burns part of the hydrogen, carbon monoxide and methane thereby raising the temperature and assisting the rapid completion of the reaction.

The process air compressor is a large consumer of energy and conventionally has been steam driven. A relatively recent development in ammonia plant design is the use of a gas turbine to drive the process air compressor and to use the turbine exhaust as combustion air in the primary reformer. The use of a gas turbine is appropriate if steam can be exported from the process to replace steam that would otherwise need to be generated on the site. A gas turbine is also normally used in those plants that use more than stoichiometric quantities of air in the secondary reformer and remove the excess nitrogen in a low temperature separation unit.

The use of high temperatures and pressures in the reforming section requires high grade materials of construction and careful design to obtain high heat transfer and optimum flow distribution to minimize equipment costs. The earlier development of ammonia plants concentrated on a high degree of conversion in the primary reformer and the recovery of heat in the flue gas. One improvement in this area is in the preheating of combustion air and the primary reformer process feed to reduce the fuel requirements. The more modern plants also operate at a much lower steam-to-carbon ratio than the traditional levels of 3.5:4.0. A reduction in process steam requirements results in considerable fuel savings, but sufficient steam has to be added to ensure that all the natural gas is reacted and that eventually the correct hydrogen-to-nitrogen ratio is obtained in the synthesis gas. Sufficient steam should be added to prevent carbon deposition in the catalyst tubes of the primary reformer that can form hot spots and reduce the life of the catalyst and tubes. One area of development to achieve a lower steam-to-carbon ratio has been the use of more active reforming catalysts that can operate with steam-to-carbon ratios of 2.5-3.0:1.0 without carbon formation.

Another method of energy conservation in the reforming section is allowing less extreme operating conditions in the primary reformer and increasing the proportion of the reaction that is carried out in the secondary reformer. In some new designs the dry gas feed is divided so that only about half the conversion takes place in the primary reformer, thereby halving the overall steam-to-carbon ratio. A large excess of air is used in the secondary reformer to complete the reaction.

Shift Conversion. Carbon dioxide is a poison for ammonia synthesis catalysts and must therefore be removed from the gases passing from the secondary reformer. This is done by converting it into the more easily removable carbon dioxide and at the same time producing more hydrogen using the water-gas shift reaction. In the early plants two stages of high temperature shift (HTS) were normally used together with some intercooling and perhaps interstage carbon dioxide removal, but it was still difficult to reduce the carbon dioxide concentration to an acceptably low level. The development of effective low temperature shift reaction (LTS) catalysts in the 1960s meant that carbon monoxide could be removed to a lower level, and today almost all ammonia processes employ both HTS and LTS stages to reduce the carbon monoxide content of the synthesis gas. The iron/chromium catalysts in the HTS section are relatively insensitive to poisons although they have the major disadvantage that they must be operated at high temperature. The LTS catalyst is much more sensitive to poisoning and most developments in this section of the ammonia plant have been to increase the life and efficiency of the LTS catalyst. One significant improvement is the use of a guard converter. The LTS catalyst absorbs poisons in a concentrated layer at the top of the catalyst bed, so a separate guard bed is inserted to act as a poison trap and enable the operator to remove the poisoned catalyst at any time in order that the main bed can be used for longer periods without changing the catalyst.

Carbon Dioxide Removal. After leaving the shift conversion stage it is necessary to remove the carbon dioxide from the synthesis gas after cooling. Most removal systems have been based on chemical absorption using monoethanolamine (MEA) solution or potassium carbonate solution. Carbon dioxide in the synthesis gas is removed in an absorption tower, and the solution is regenerated by heating at a lower pressure to remove CO_2 before being recirculated to the system. This part of the process requires energy to regenerate the scrubbing solution. This energy can be provided as low pressure steam or more recently by recovering heat from the process gas leaving the LT shift section. In order to improve energy efficiency, systems have been developed to improve the absorption capacity of the circulating solution by operating at higher concentrations.

Unfortunately in the case of MEA, higher concentrations mean increased corrosion; this is being overcome by the addition of corrosion inhibitors. Other developments include the use of non-corrosive methyl-diethanolamine (MDEA) that can be regenerated more cheaply than MEA.

Hot potassium carbonate solutions also have a high capacity for removing CO_2, but unfortunately the absorption rates are lower than ethanolamine solutions. However, the solution properties have been improved by the addition of diethanolamine and corrosion inhibitors. Two of the best known potassium carbonate systems are the Benfield and the Giammarco-Vetrocoke processes. A recent advance in the Benfield process is the Loheat process that claims significant savings in energy. The low energy requirements of the system are the result of recovering sensible heat supplied to the solution in the regenerator as flash steam by reducing the pressure of the solution after steam regeneration. This steam is compressed and recirculated to the regenerator.

Carbon dioxide can also be removed from the synthesis gas by physical absorption, and the advantage of this process is that air rather than steam can be used for regeneration, which can result in energy savings. A disadvantage is the high cost of the solution. Physical solvents that are being used successfully include Selexol, Fluor solvent, Purisol and Sepasolv MBE.

Carbon dioxide recovered from ammonia plants has a number of possible uses, and often one of the most convenient is as a feedstock together with ammonia for the production of urea.

Methanation. The gas leaving the carbon dioxide absorption step contains small quantities of carbon monoxide and carbon dioxide that must be removed before the ammonia synthesis step because even these small quantities would decrease the activity of the ammonia synthesis catalyst and cause deposition of ammonium carbamate. The methanation reaction that converts carbon monoxide and carbon dioxide into methane and water is the reverse of the reformer reactions. Until the introduction of the LT shift catalyst in the early 1960s carbon monoxide was removed using either a copper liquor wash or a liquid nitrogen wash, and traces of carbon dioxide were removed with an alkali wash. A traditional nickel catalyst is now used and most of the research work on this stage has been directed to improving the selectivity and activity of the methanation catalyst.

Ammonia Synthesis. Before passing to the synthesis loop it is necessary to clean-up the outlet gases from the methanation stage by cooling and drying. The gases are cooled by heat exchange and finally by refrigeration. Condensed water in the cooled gas is removed and the chilled gas is fed to a drier containing a solid desiccant or molecular sieves to remove ammonia, residual carbon dioxide and water. The dry gas is now mixed with recycled gas and circulated to a preheater after which it enters the ammonia converter. The gases from the converter are cooled by heat exchange with the inlet gases and then refrigerated to recover ammonia. Inerts like argon, helium and methane do not dissolve sufficiently in the recovered ammonia and have to be purged from the system. This purge gas contains about 60% hydrogen, and when natural gas is cheap it can be used as a fuel. When natural gas is expensive it is usual to install a hydrogen recovery system. Hydrogen can be recovered either by membrane separators or by low-temperature separation. The hydrogen-rich stream is returned to the circulating system and the concentrated purge gas is available for use as a fuel.

The basic chemical reaction for the production of ammonia is simple, but in practice it has to take place at high temperature and pressure enhanced by a catalyst:

$$3H_2 + N_2 = 2NH_3$$

The equilibrium between the product ammonia and the reactants depends on temperature and pressure. The reaction is exothermic, and in order to prevent the temperature in the catalyst bed from rising to a point where conversion would be unacceptable, the reactor has to be cooled. The normal optimum operating pressure range for ammonia synthesis using conventional iron-based catalysts is between 150 and 300 atm, but catalysts have recently been developed that allow some of the new low energy processes to work with advantage at lower pressures: for example, at 70-80 atm.

There are many different types of reactor designs available; the characteristic features are the gas flows through the catalyst bed, the method of temperature control in the bed and the recovery of the heat of reaction. Ammonia converters can be basically classified according to the method by which the reacting synthesis gas is cooled. In a quench converter, cooling is achieved by injecting cool fresh synthesis gas directly into the system, whereas in indirectly cooled converters the reacting gases are cooled in a heat exchanger system. The new low-energy ammonia plants mainly use two bed converters with intercoolers. One of the advantages of the system is that steam can be generated in the synthesis loop.

The New Energy-Saving Ammonia Processes

The production of ammonia involves a succession of high temperature gas-phase reactions from which a substantial amount of heat must be recovered. The major objective in recent ammonia plant design has been to

196

recover this heat in the most efficient way to provide the power requirements for the operation of the plant under normal conditions. The evolution of ammonia plant design by the major engineering/operating companies has produced a variety of energy recovery and plant utilities systems. Although the configuration of heat exchangers, drives, operating conditions, etc., can vary significantly from one design to another, the overall integrated effect in all the leading designs has been to produce a plant reliable and capable of high utilization rates and with low energy consumption in the range 25-28 MMBtu per metric ton of ammonia. These plants are generally comparable in overall performance, and one of them that first came into service in 1985 is described briefly below and in Figure 6.10.

The ICI AMV Ammonia Process. Natural gas is mixed with recycled hydrogen, heated and desulphurized. It is then cooled by preheating the feed to the desulphurizer before passing to a feed-gas saturator where it is contacted with circulating hot process condensate.

The feed gas from the saturator is mixed with a further quantity of steam to give a steam-carbon ratio of the order of 3:1, preheated in the reformer flue gas duct and reformed at 700°C and 400-500 psig. The gas mixture is then fed to a secondary reformer for further reforming with an excess of unheated process air. The secondary reformer operates at a temperature of 850-950°C. The reformed gas is cooled by generating superheated high-pressure steam and then shifted in high- and low-temperature shift converters. The cooled gas from the LT shift converter is treated to remove CO_2, compressed and methanated before being fed to the ammonia synthesis loop, which operates at 1000-1200 psig. Circulating gas from the ammonia synthesis loop is mixed with the dried synthesis gas and fed to a circulator and then heated before passing over a new ICI low-pressure synthesis catalyst to produce ammonia. After cooling, ammonia is removed from the gas by mechanical refrigeration.

Inerts and excess nitrogen are removed by taking a purge from the circulator delivery and treating it in a hydrogen recovery unit. The recovered hydrogen is recycled to the circulator suction.

Future Trends in Ammonia Plant Design

With energy consumption now approaching a practical limitation, increased attention is being given saving investment costs in plant design by rationalization of the utilities system and saving the cost of equipment by operating under less severe conditions. Improved catalysts will probably in time permit the reduction of temperatures in the reforming, shift and synthesis stages. There is already a trend to increase the flexibility of operation by moving from steam turbines to more efficient

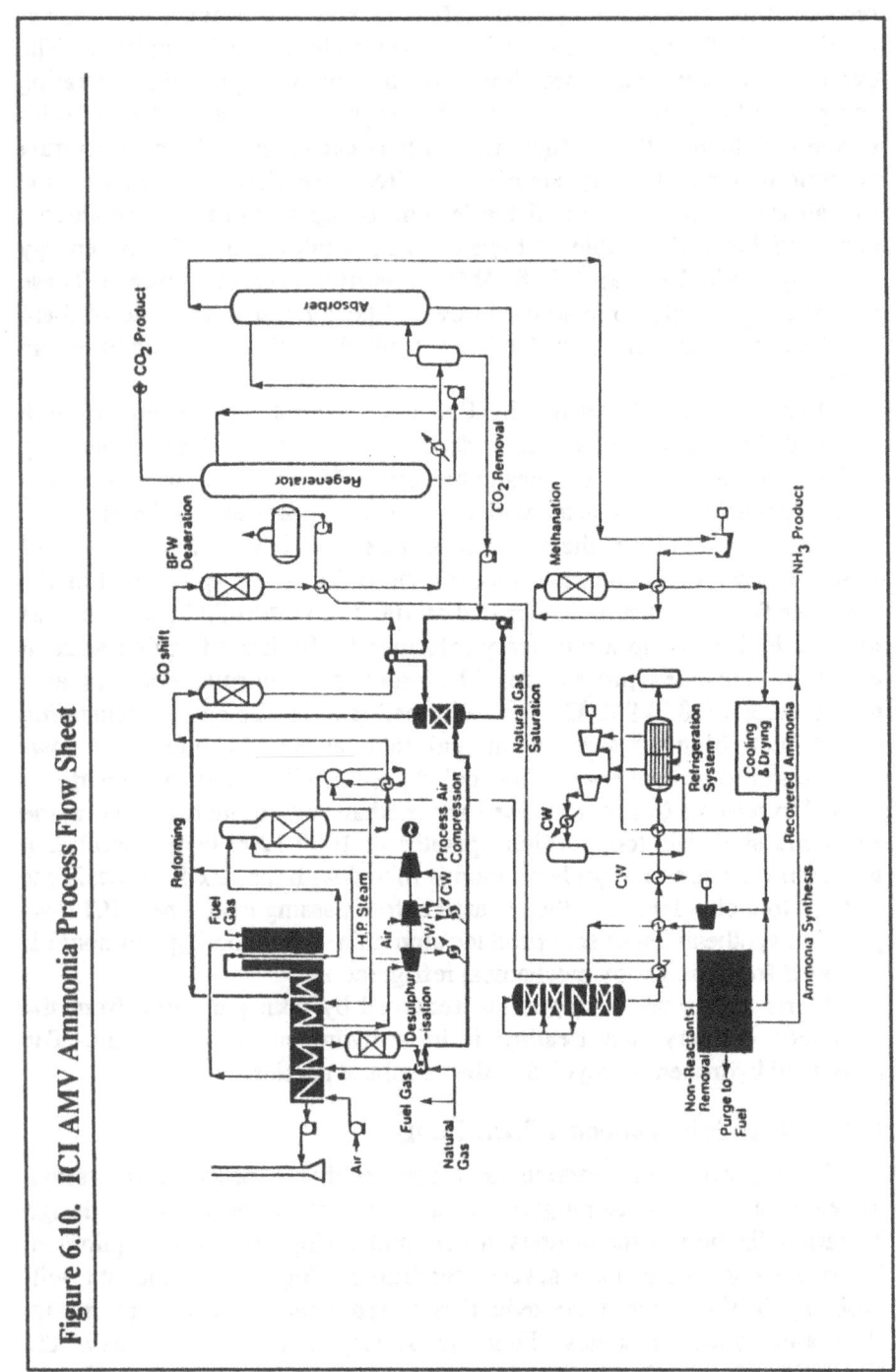

Figure 6.10. ICI AMV Ammonia Process Flow Sheet

electric motor drives for small- and medium-sized duties. In some cases savings can be achieved by using gas turbine drives. The increasing use of computers to simulate and control the process will ensure more stable and efficient operating conditions.

The concept that larger plants are necessarily the most economic is being challenged by some designers who are developing new smaller scale plants that can reportedly compete with larger plants in terms of thermal efficiency and unit investment cost. The newest is the ICI LCA plant; two of these, each of 450-tpd capacity, were commissioned in the UK in 1988. These plants contain a number of novel features, such as a fundamental simplification of the steam and power cycle in which the "core" unit containing the key operations for making ammonia is separated from the utilities area containing the steam and power systems. Complex heat recovery systems are eliminated from the core unit, which is further simplified by the use of a gas-heated reformer in which the primary reformer receives heat directly from the process gas exiting the secondary reformer. This eliminates the large burner structure and high-pressure steam system of a conventional plant. Based on new catalysts, the process employs a low-pressure synthesis loop, and other novel features include a single-shift conversion stage and a pressure-swing adsorption system.

It is claimed that by reducing the weight of the main plant items and using prefabricated modules, the diseconomies of working at a smaller scale are reduced and can be outweighed by the reduced erection and start-up times and by the benefits of phased investments. It seems likely, however, that many of the beneficial features of the LCA process will be introduced into new large plants in the not too distant future.

The Economics of Ammonia and Nitrogen Fertilizer Production

General

The two main factors in determining the cost of ammonia and nitrogen fertilizer production costs are feedstock costs and investment costs (capital charges), and these can vary significantly for different locations. Together these two components make up about 90-95% of the total cost of production. Sometimes raw materials can be available very inexpensively, but this advantage can be offset by higher investment costs if plants have to be built in remote locations and bear the costs of extensive infrastructure.

There are two aspects of the comparative economics of nitrogen fertilizer production that have to be examined carefully. One of these is the total cost of production, including the cost of capital, which is sometimes referred to as the "realization price." This is the price to justify new investment taking into account the cost of raw materials, the required return on investment and the capacity and utilization of the plant.

Another important parameter in comparing the economics of different plants is the cash cost of production, which for ammonia basically reflects the cost of feedstock and can vary significantly from project to project. This is particularly important in an industry where sales prices fluctuate considerably and where on some occasions prices may fall to a level that would be insufficient to allow a company to meet its day-to-day operating costs.

Energy Requirements and Costs

The cost of energy in the form of fuel and feedstock is usually the most important component of ammonia and nitrogen fertilizer production costs. This is clearly illustrated by information obtained from the surveys carried out periodically by The Fertilizer Institute (TFI) of the North American fertilizer industry. Some information from the survey for 1987 that covered more than thirty US and Canadian plants is referred to in Tables 6.5 and 6.6. The TFI study is useful as it relates to actual operating experience rather than engineering companies' estimates of "battery limits" consumptions achieved under optimum conditions. It also aggregates all types of energy into equivalent units.

Unfortunately, energy needs in other cases are not always presented on a consistent basis, often because of the difficulties of defining the boundaries for energy use when there may be both imports and exports of energy out of the system. Sometimes emphasis and comparisons are made only of process energy needs, and often total energy needs are expressed as a mixture of different forms of energy and units, which makes comparisons difficult. Under normal operating conditions, total energy needs, including energy for operating offsites and handling facilities, for shutdowns and start-ups and due to maloperation, etc., may exceed the guaranteed battery limit figures by 10-15%.

Although the overall energy usage may appear to be on the high side compared with the consumptions now quoted for new plants, this is because about 70% of US plants operating today were constructed before 1975 when energy requirements exceeded 40 MMBtu/metric ton of ammonia. A further 25% of existing capacity was constructed between 1975 and 1980 when energy consumption averaged about 35 MMBtu/metric ton of ammonia. A few plants have recently been refurbished to improve output and conserve energy, but because of the relatively low gas prices in the US in the last few years these upgrades have not been widespread.

200

Table 6.5. Average Energy Consumption in Anhydrous Ammonia Production[*]

	Natural Gas	Imported Electricity	Steam	Total
Reciprocating Plants[a]				
Total Energy Use				
Feedstock Energy	24,074			24,074
Reformer Process Energy	8,272	419		8,691
Other Process Energy	5,070	3,631	(396)	8,305
Total	37,416	4,050	(396)	41,070
Centrifugal Plants[b]				
Total Energy Use				
Feedstock Energy	23,929			23,929
Reformer Process Energy	12,672	4		12,676
Other Process Energy	2,373	712	(360)	2,725
Total	38,974	716	(360)	39,330

[*]1000 BTUs per metric ton of ammonia
[a]Based on 1,295,000 metric tons production
[b]Based on 8,384,000 metric tons production

Naturally, as energy prices have increased, a great deal of attention has been focused on reducing the energy needs of ammonia plants. The overall effect of process development on reducing energy consumption is shown in Figure 6.9. Energy consumption per unit of product has fallen from about 75 MMBtu/metric ton in the early 1950s to less than 30 MMBtus for plants built after 1985. Most of the major ammonia process designers now offer low-energy designs that are likely to result in *total* energy consumptions in practice within the range 28-32 MMBtu/metric ton or even less. Such designs are not the result of technical "breakthroughs," but more the integration of a number of energy conservation features. The introduction of these new features does increase the plant costs and also makes the plant more complex to operate. When gas prices are less than US$1.0/MMBtu and likely to stay so, the benefits of some of the new energy saving design features become marginal. Further major decreases in energy consumption may no longer be achievable because energy consumption for the most recent process schemes are close to the theoretical and practical levels.

Table 6.6. Natural Gas Costs for US Ammonia Production

Calendar Year Production	US$/MMBTU (Average) Cost	US$/Metric Ton of Ammonia	% of Total
1987	1.71	58.63	64.9
1986	1.83	64.35	65.3
1985	2.65	95.14	76.0
1984	2.68	95.84	75.8
1983	2.34	91.24	72.0
1982	2.40	95.17	72.1
1981	2.33	93.51	72.6
1980	1.96	78.04	69.4
1979	1.62	65.03	66.8
1978	1.34	55.19	61.7
1977	1.14	46.37	57.2
1975	0.62	24.89	46.4
1973	0.36	15.30	45.6
1970	0.28	12.08	41.5

Total Production Costs for Ammonia and Urea

Total production costs for ammonia and urea based on natural gas are presented in Figures 6.11 and 6.12 and indicate clearly how these costs vary for differing gas and investment costs. The model used to prepare the total production costs as illustrated is typical of those used for appraising and comparing the viability of new nitrogen fertilizer projects. Although in practice a considerably longer life is likely, for appraisal purposes a life of 15-17 years is normally assumed. In this case a 15-year life has been taken.

Investment Costs. The investment costs in 1988 US dollars for a 1000-tpd ammonia plant including offsites on a greenfield site in a developed location with existing infrastructure would be about US$150-200 million. If a matching urea plant of 1670-tpd capacity is included the total investment cost would be increased by US$30-40 million. In developing areas the cost of infrastructure can add considerably to total costs, particularly if it is necessary to build a railroad, township, port facilities, etc. In remote locations the total investment cost for an ammonia urea complex could be US$500 million or more. A utilization of 90% is assumed, which is typical for a plant in a developed country. In 1988 the average utilization for all world ammonia capacity was 83%. Utilization

Figure 6.11. Realization Price of Ammonia Production

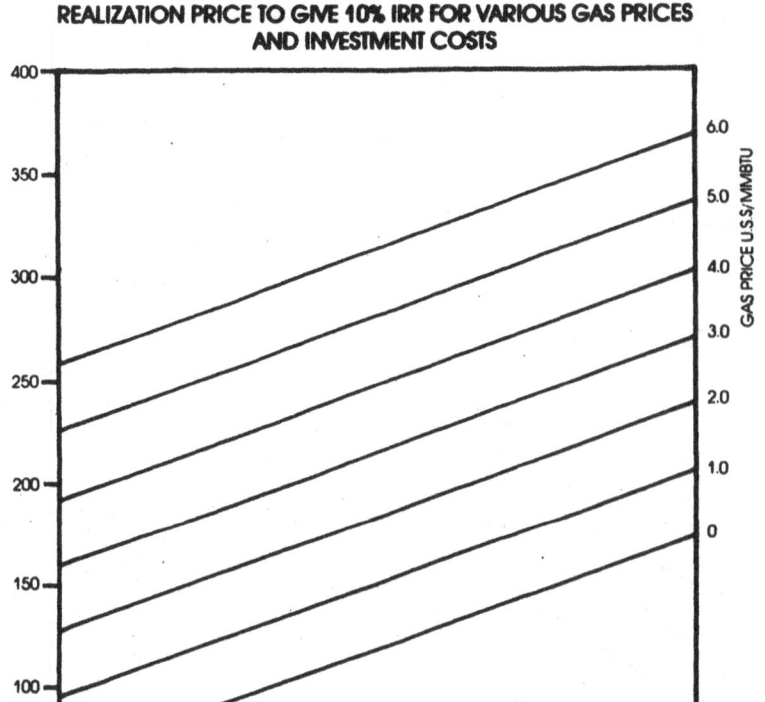

AMMONIA PRODUCTION — 1350 T.P.D.
REALIZATION PRICE TO GIVE 10% IRR FOR VARIOUS GAS PRICES
AND INVESTMENT COSTS

Figure 6.12.   Realization Price of Urea Production

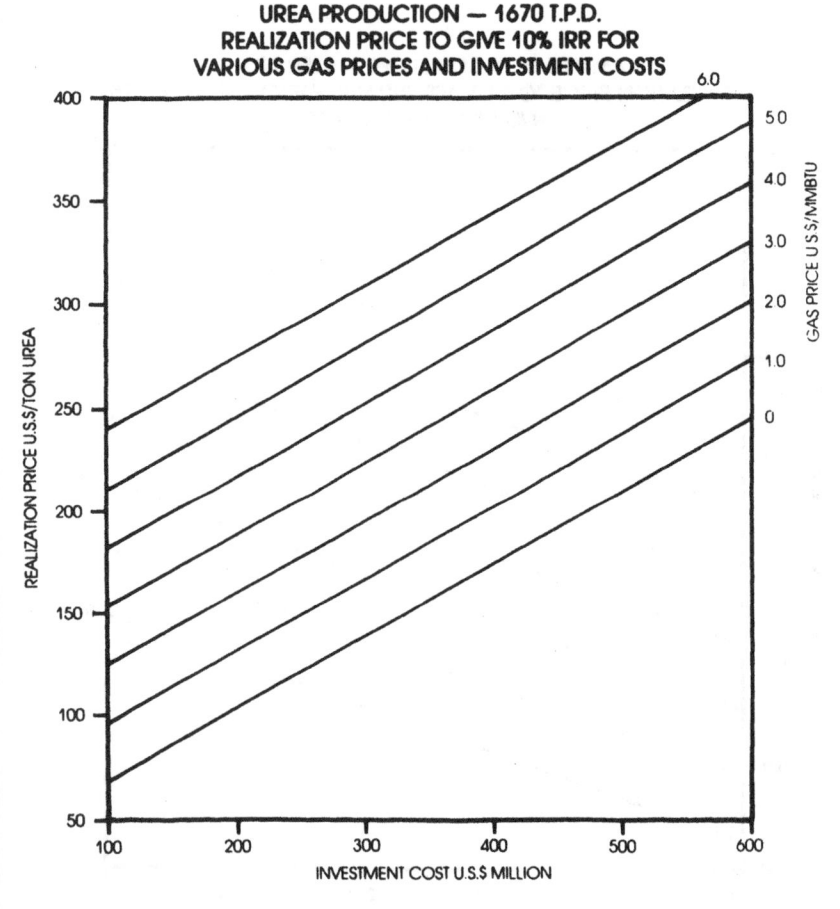

UREA PRODUCTION — 1670 T.P.D.
REALIZATION PRICE TO GIVE 10% IRR FOR
VARIOUS GAS PRICES AND INVESTMENT COSTS

Figure 6.13. Sensitivity of Realization Price to Gas Costs for Different Ammonia Scales of Production

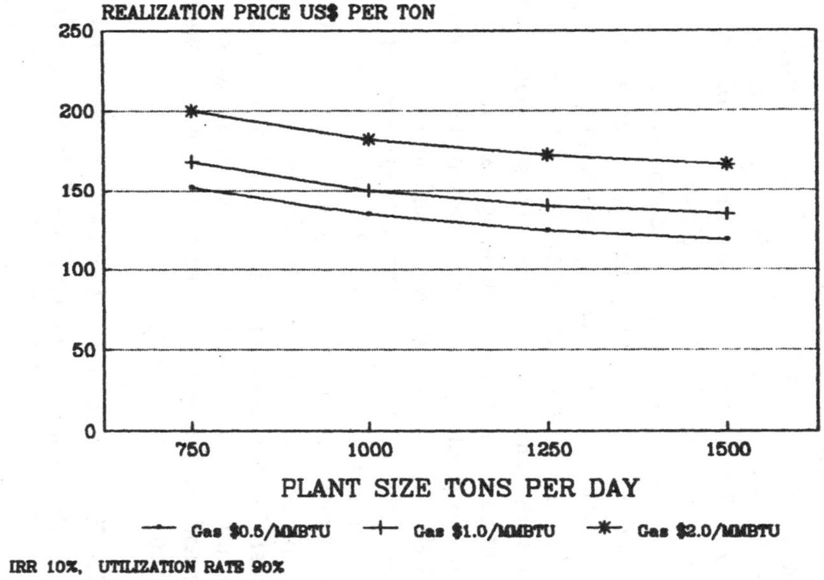

REALIZATION PRICE US$ PER TON

PLANT SIZE TONS PER DAY

— Gas $0.5/MMBTU   + Gas $1.0/MMBTU   * Gas $2.0/MMBTU

IRR 10%, UTILIZATION RATE 90%

rate affects the capital charge component of total production cost, so it is much more important in those cases where investment costs are high. Because the relationship between investment cost and plant size is not linear, there is normally an economic advantage in building larger size plants and this is shown in Figure 6.13. In the case of ammonia the capital-related costs per ton for a 1350-tpd plant compared with a 1000-tpd plant on a developed site is about US$16 less. For a developing site the comparable difference could be considerably more. The penalties of reduced size apply mainly to a situation of a single plant on one site. Two smaller plants sharing the same offsites and infrastructure as a larger plant of the same overall capacity may not be at a significant disadvantage comparing investments and could in fact have more flexibility with regard to operation. This would be the case for some of the new medium-size ammonia plants such as the ICI LCA plants.

Variable Costs. A total energy consumption of 32 MMBtu/ton of ammonia and 28 MMBtu/ton of urea has been assumed. These figures would be typical of average energy consumptions for plants coming on stream before 1988. It is appreciated that these figures may seem high compared with battery limit energy consumptions quoted by process vendors, but audits of existing operating plants and reference to TFI data indicate that in practice total energy consumptions are somewhat higher. Other variable costs are relatively small and in the case of ammonia amount to about US$5.0/ton. The variable cost figure of US$18/ton assumed for urea includes US$13/ton for bags.

Fixed Costs. Labor and supervision costs have been estimated at US$2.5 million/year for a 1000-tpd ammonia plant and US$4.5 million/year for the ammonia plant and US$2.8 million/year for the ammonia urea complex. Capital-related costs to cover maintenance, insurance and local taxes have been taken as 4% of plant investment and 2% of the cost of infrastructure.

Table 6.7. Relative Investment and Energy Needs for
Producing Ammonia Using Different Feedstocks

Type of Plant	Ratio of Investment Cost	Ratio of Energy Consumption
Natural Gas	1.00	1.00
Naphtha	1.15	1.08
Fuel Oil	1.60	1.15
Coal	2.00	1.40

The Cost of Producing Ammonia with Different Feedstocks

Although natural gas is currently the most commonly used feedstock for ammonia production and is expected to remain so, there may be occasions depending on relative energy pricing when other feedstocks may be competitive. The most important factors in determining the production costs are the relative investment and energy costs shown in Table 6.7 for a developed site.

Although there is much information available on the investment and production costs of gas-based processes because very few other types of processes have been built in recent years, information on comparable up-to-date processes for the other feedstocks is not so readily available. It has been assumed that certain improvements in energy conservation have taken place relative to what has been achieved in gas-based plants. Equivalent energy prices which have been used in the comparison are given in Table 6.8.

The advantage of natural gas as a feedstock can be clearly understood from Figure 6.14 because of the reduced investment cost and energy needs. The comparison is made on the basis of a 10% IRR. At higher rates of return the comparison favors natural gas even more. It is only when gas prices become very high, (say, at least, US$5/MMBtu) that coal becomes competitive.

Table 6.8. Equivalent Energy Prices for Different Fuels

Natural Gas US$/MMBTU	Naphtha US$/Ton	Fuel Oil No. 6 US$/Ton	Coal US$/Ton	Crude Oil US$/BBL
1.0	44.8	41.9	25.1	7.1
2.0	89.6	83.8	50.2	14.2
3.0	134.4	125.7	75.3	21.4
4.0	179.2	167.6	100.4	28.5
5.0	224.0	209.5	125.5	35.6
6.0	268.8	251.4	150.6	42.6

Figure 6.14. Ammonia Realization Price vs. Feedstock Cost

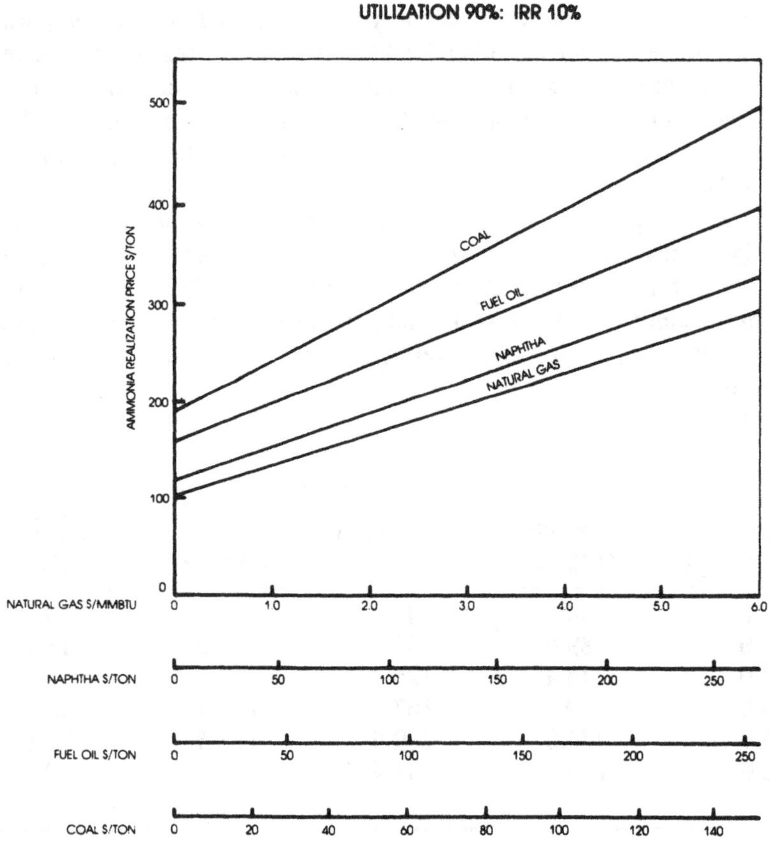

UTILIZATION 90%: IRR 10%

The Economic Cost of Natural Gas
for Ammonia Production

There are many countries with reserves of natural gas that could be used for ammonia production. In some cases ammonia projects are not realized because the investment cost of recovering the gas and establishing a nitrogen fertilizer complex is too expensive. In other cases one of the most important factors determining the feasibility of ammonia or urea production will be the economic (opportunity) cost of the particular gas resource available or to be made available for such production. It is difficult to generalize abut the economic value of gas because it varies from location to location depending on the size of the resource and on the alternative (opportunity) uses of the gas if it were not to be used for fertilizer production. If the gas can be used for oil substitution in power stations, the economic value of the gas would be linked to the value of oil. However, in many countries, particularly developing countries, there are many occasions where this fuel oil substitution alternative is not available.

Alternative uses of gas include the nitrogen fertilizer and petrochemical industries such as methanol or liquefied natural gas (LNG) for overseas markets. Generally, because of their similar investment and processing costs the economics of methanol and ammonia are similar. Where deposits of gas are small, ammonia and urea plants are often the most attractive proposition, particularly if some of the fertilizer can be consumed locally.

Much consideration has been given to the liquefaction and transportation of natural gas particularly during the period when oil prices rose to more than $30/BBL during the early 1980s. The cost of liquefaction and transport of natural gas is expensive, as can be seen from Figure 6.15, and the opportunity cost for gas used for LNG manufacture will be the net-back price after subtracting the liquefaction and transport costs from the corresponding oil price.

When oil is $30/BBL there is generally going to be a good case for LNG exports with net-backs equivalent to between $1.0-2.0/MMBtu. When oil is priced at $20/BBL the cost of transport becomes very important and there are few opportunities for LNG. Because of the very large investments needed and the uncertainty of future oil prices, exports of LNG have not developed as was expected. Generally these factors have helped to increase the availability and reduce the price of gas for ammonia manufacture, which in many energy rich countries is available at $1.0/MMBtu or less. In many cases, particularly where gas is being flared and has no apparent alternative immediate use, the opportunity cost of gas is basically that of collection and sweetening.

Figure 6.15. Cost of Liquefaction and Transportation of Natural Gas

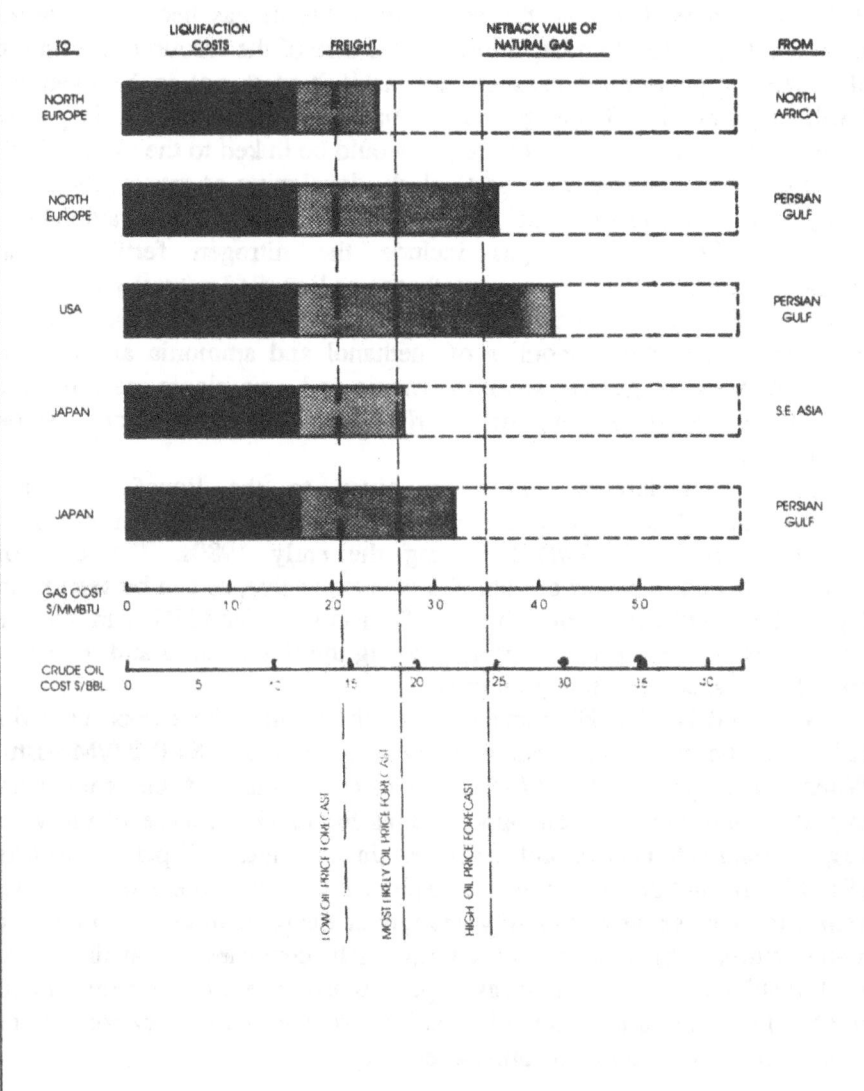

The production of ammonia and nitrogen fertilizers will continue to remain one of the best prospects for natural gas exploitation for many developing countries both for domestic use and for exports. Although the opportunity cost of gas will set the lower level of gas price on which economic returns are calculated, in many cases the financial prices are often on a higher scale or variable scale that is related to the price of ammonia or urea.

The Outlook for Natural Gas Prices
for Ammonia Production

Natural gas prices for ammonia production vary widely, from well under US$1.0/MMBtu in the energy rich countries to more than US$5.0/MMBtu in other countries with more limited energy resources. Although generally ammonia export prices have followed the same trend as world oil prices, it is only in Western Europe that prices of natural gas are related to and reflect other energy prices. In many other countries the gas and oil markets generally appear to operate independently. It is estimated that about 40 million tons (N) of ammonia capacity out of 80 million tons (N) of total world gas-based capacity had a gas price of about US$1.0/MMBtu. About 25 million tons (N) of ammonia capacity had a gas price between US$1.0-3.0/MMBtu, and the remaining world capacity operated about this range.

North America. As indicated in Table 6.6 average natural gas prices for the US ammonia industry have declined steadily between 1984 and 1987 from US$2.68 to US$1.73/MMBtu. Generally the gas and oil markets have operated relatively independently in the US. In a balanced market one might expect that gas as a premium fuel would command a slightly higher price than oil, but the strong competition among gas producers and transmission companies has resulted in natural gas prices being substantially less than oil prices on an equivalent energy basis. The nature of the US gas industry has changed since mid-1986, with a much greater emphasis on drilling for gas; in 1987 there was only one quarter of the rigs in use compared with 1981. It seems likely that in the next few years the availability of gas in the US will decline and that prices will increase to approach parity with fuel oil prices. The uncertainty of the gas market in the US and the fact that neighbors such as Canada and Caribbean countries will have much cheaper gas available may constrain further investment in new ammonia capacity in the US.

Canada has large reserves of natural gas mainly in Alberta and British Columbia. Deregulation of energy prices in Canada in 1986 had the impact of reducing gas prices, and in 1987 the price of gas available to new ammonia producers in Alberta was in the range US$1.0-2.0/MMBtu.

211

Prices of gas in Canada are expected to remain lower than gas prices in the US, and imports of ammonia from Canada to meet a growing deficit in the US are likely to increase.

Western Europe. Many countries in Western Europe have a gas price related to fuel oil prices so that ammonia prices follow very closely the price of fuel oil. In 1987 gas prices in Western Europe averaged about US$2.4/MMBtu, and these prices are expected to increase slowly in line with forecasts for increasing oil prices. One exception is in Ireland where the gas resulting from a long-term contract with the Marathon offshore gasfield is significantly lower. Because of relatively high gas prices no major expansions in ammonia capacity are expected although plants supplying local markets may be replaced.

Africa. This region has several sources of cheap natural gas for ammonia production; for example Nigeria, Libya and possibly in the future Tanzania, Mozambique and Cameroon all have prices below US$0.5/MMBtu. The main constraint to investment in Africa apart from Egypt and Nigeria is the lack of local markets in the high cost of infrastructure. Low gas prices are therefore necessary to entice new investments for export-based projects.

Near East. The Near East has developed since the late 1970s as a major producer and exporter of nitrogen fertilizers. Gas prices are approximately US$0.5 -1.0/MMBtu and will probably stay low. With large gas reserves in Iran, Qatar, Saudi Arabia and elsewhere, the region is expected to expand its industry significantly in the next decade. With considerable infrastructure now in place most new plants will go on existing sites, thus significantly reducing investment costs.

South East Asia. In Pakistan and Bangladesh the cost of natural gas to the fertilizer industry is about US$1.0/MMBtu. In India natural gas prices to fertilizer producers vary between about US$3 to US$6/MMBtu depending on plant location and transmission costs. In this situation naphtha and fuel oil can compete with gas in some locations. In Indonesia and Malaysia gas prices are about US$1.0/MMBtu or slightly less. Gas prices for new plants will be increased to US$1.5/MMBtu or more.

Eastern Europe. It is difficult to put a value or price on gas in Eastern Europe, as ammonia is exported to earn foreign exchange even at very low prices. Based on these prices over the last few years, it is unlikely that the net-back price for gas has been more than US$1.0/MMBtu. In China gas prices are not uniform and do not appear to be set according to any established criteria. The government sets the price for each plant.

The Outlook for the
World Nitrogen Fertilizer Industry

World Nitrogen Demand

World and regional nitrogen demand for the fertilizer industry and for industrial use through the year 2000 are summarized in Table 6.9. These projections are based mainly on those of the World Bank/FAO/UNIDO Fertilizer Working Group but also take into account projections made by the World Bank using its Integrated Agricultural/Fertilizer model.

The growth rate for total nitrogen is expected to average about 2.2% p.a. through the year 2000. For fertilizer growth the rate is expected to average about 2.3% p.a. and for industrial nitrogen about 1.5% p.a. Very little growth is expected in the industrial countries of North America, Western Europe and Japan where application rates are already near economic optimum. There could in fact be a slight decline in nitrogen fertilizer consumption due to pressure from environmentalists concerned about the leaching of nitrogen fertilizers into drinking water.

The highest rate of growth will take place in the developing countries, mainly South East Asia. Nitrogen application rates in many of these countries are still relatively small and many of these countries have the potential and the need to increase fertilizer use to meet the growing food requirements of an expanding population.

The consumption of nitrogen fertilizers in centrally planned economies is expected to increase at about 2.3% p.a. China has already achieved a very high application rate of about 138 Kg/Ha, more than thrice the rate of India (39Kg/Ha), and the rate of increase in consumption is now expected to slow down somewhat. Many countries in Eastern Europe already have relatively high application rates, and the biggest increase in demand will probably take place in the USSR. This country is a major

Table 6.9.   Nitrogen Demand in 1999/2000 Million Tons and Average Annual Growth Rates for 1987/88-2000/01

Region	Fertilizer		Industrial		Total	
	Mill. Tons	%	Mill. Tons	%	Mill. Tons	%
Developed M.E.	24.3	0.3	7.6	0.9	31.9	0.5
Developing M.E.	30.8	4.3	0.9	3.3	31.7	4.3
Centrally Pl.E.	45.0	2.3	3.3	2.6	48.3	2.3
World Total	101.1	2.3	11.8	1.5	111.9	2.2

importer of food and will undoubtedly try to increase its domestic food production based on an abundant domestic supply of nitrogen fertilizers.

The comparisons in Table 6.10 show how far the developing market countries lag behind both the developed and centrally planned economies in nitrogen fertilizer use. However, the increasing use of nitrogen fertilizers in socialist Asia (mainly China) has been spectacular, growing on average at more than 20% per year over the last thirty years or so. The use of nitrogen fertilizers has enabled China to increase its yield and food production more than four times during a period when the availability of arable land has even declined slightly.

The Supply of Nitrogen Fertilizers

A world-wide massive build-up in ammonia and nitrogen fertilizer capacity particularly in the Soviet Union at the end of the 1970s produced a large surplus that has persisted throughout the 1980s and seems likely to continue until the early 1990s. Thereafter there will be an average annual need for new capacity of at least 2-2.5 million tons N, equivalent to increasing demand. A significant amount of new capacity will also be required to replace obsolete or worn-out plants.

With good and routine maintenance and regular replacement of certain items there seems no reason to doubt that the life of a large ammonia plant can be extended to thirty years or so. However, more than half of the existing capacity of about 117 million tons or so was built before 1970 and plant closures that have been averaging more than one million tons per year are likely to double during the 1990s.

Location of New Ammonia Capacity

In considering the location of new nitrogen fertilizer plants it is important to look at the markets for their products. It is also necessary to

Table 6.10. Consumption Rates for Nitrogen Fertilizers, 1986

Region	Kg N per Ha of Agricultural Area	Kg N per Ha of Arable Land and Permanent Crops	Per Caput
Developed M.E.	18.2	58.4	28.2
Developing M.E.	8.6	27.3	7.4
Eastern Europe & USSR	24.0	57.7	40.5
Socialist Asia	26.0	129.5	12.4

Source: *FAO Fertilizer Yearbook 1987.*

realize that there are basically two markets to consider, one for ammonia and one for urea. Ammonia is marketed mainly in the US and Western Europe where it is used to produce ammonium nitrate and ammonium phosphates. Other countries that import ammonia are the major phosphate-producing countries of Morocco, Tunisia and Jordan. The best locations for new ammonia capacity for the export market will continue to be in Eastern Europe both from the Baltic and Black sea ports and also from the Caribbean countries such as Trinidad and Venezuela that are well endowed with cheap natural gas.

Urea is the preferred fertilizer for wet paddy, so the main market will be the expanding market of the Far East. It seems likely therefore that there will be further investments in urea plants in China, India, Indonesia and possibly in Burma and Bangladesh. There are also several locations in Africa such as Tanzania, Mozambique, Nigeria and Libya where consideration is being given to new nitrogen fertilizer plants. The Near East will continue to attract new investment because of its cheap gas and convenient location for the export markets of the Far East.

It is estimated that about 60% of all new ammonia capacity between 1990 and 2000 will be built in the developing market economies, about 30% in the centrally planned economies and only about 10% in the developed market economies.

The Impact of Biotechnology on
Nitrogen Fertilizer Use

The question is often raised whether biotechnology and genetic engineering will have an impact on future nitrogen fertilizer use and hence on the needs for new ammonia plant capacity. Considerable progress is being made in crop genetics where new varieties are being developed that are more resistant to diseases and insects and have higher yields. However these plants, like current varieties, still need chemical fertilizers, and developments in this area are not likely to affect nitrogen fertilizer use in the short or medium term. In some cases where future varieties make it possible to use crops on marginal land, nitrogen fertilizer use could actually be increased.

Research is being carried out to create new varieties of plants that will permit the use of less fertilizer by fixing atmospheric nitrogen: for example, microbial nitrogen fixation to combat the Rhizobia that live in legumes like beans and peas. The study of microfixing organisms is probably one of the most active areas in present genetic research.

Current research suggests that soybean nodules and legumes can probably be improved, but the impact on fertilizer use is not likely to be significant. Calculations on nodulating cereals suggest that because the

215

biological-fixing process is energy demanding it would be more profitable for the farmer to use fertilizer to maximize yields rather than use plants that would have a nitrogen-fixing property but would result in a lower yield. However, it may be that in some developing countries such nitrogen-fixing plants could be of benefit.

Another long-term possibility may be the genetic engineering of nitrogen-fixing plants that have the ability to fix nitrogen in the leaves rather than the roots. In this case the leaves would usually have sufficient energy to fix the nitrogen, but because the presence of oxygen makes the nitrogen-fixing enzymes inactive, such a long-term solution may not be possible unless enzymes can be developed that are oxygen tolerant.

Much effort has gone into nitrogen-fixing enzymes, but there are still many fundamental problems to be solved before the technique can be widely used and have a significant impact on fertilizer use. The techniques are still very much in the laboratory stage of experimentation. Even if major breakthroughs occur it would take many years for them to be widely applied in practical farming and thus significantly reduce the need for chemical fertilizers. It is unlikely that the developments will be so rapid as to make obsolete new investments in ammonia plants.

Environmental Constraints to
Nitrogen Fertilizer Use

Environmental constraints to nitrogen fertilizer production come mainly in the form of regulations regarding the emissions of noxious gases and dusts and from the handling and storage of potentially dangerous materials such as ammonia and ammonium nitrate. Although they add considerably to the cost of producing nitrogen fertilizers, these constraints can be overcome by additional processing, improved design of plants and attention to safety precautions.

There are two environmental problems associated with the use of nitrogen fertilizers that are causing growing concern and that are already having an impact on nitrogen fertilizer consumption. At this stage the most serious is the leaching, or run-off, of nitrates. In many cases the increase in the nitrate level of drinking water is being associated with the increasing use of chemical fertilizers particularly in those developed countries where fertilizer application rates are high.

In several countries in the European Economic Community (EEC) nitrate levels in drinking water already exceed recommended levels, and in order to reduce these levels new EEC directives could seriously limit the use of nitrogen fertilizers and even result in an overall fall in consumption in this region. Legislation has also been effected or is under consideration in certain states in the US to monitor the quality of drinking water.

216

Undoubtedly these restrictions on the use of nitrogen fertilizer could in certain areas have a major impact on fertilizer use. Experimental work indicates that when fertilizer nutrients are applied in a balanced manner, nitrate leaching can be minimized.

Another problem now being related to nitrogen fertilizers but more difficult to evaluate is the relationship between increased fertilizer use and global climate change. It is believed that nitrous oxides released to the atmosphere through the denitrification of nitrogen fertilizers in the soil may lead to reactions in the stratosphere that assist in the destruction of the ozone layer. About 14% of nitrous oxide emissions are believed to come from fertilizer use, but it is likely that this figure might be reduced through more efficient placement of fertilizers and the use of nitrification inhibitors.

Conclusion

The availability of low-cost natural gas in the 1960s and 1970s had a major impact on the development of the nitrogen fertilizer industry as it simplified and reduced the cost of producing ammonia. This helped to sustain the "green revolution" that resulted from the introduction of high-yield varieties of cereals that responded economically to much higher application rates of nitrogen fertilizers. The continuing availability of low-cost natural gas in most regions and particularly in developing countries has ensured that nitrogen fertilizers are freely available to farmers throughout the world.

In order to meet future food needs, nitrogen fertilizer consumption will need to increase at about 2-2.5% on average per year. Although new ammonia plants are reaching their theoretical limit with regard to energy and hence natural gas use, there are many existing older plants where energy consumption is 30% or more than required in the new technology. As many older plants are replaced or revamped it is possible that the world need for natural gas for ammonia production might increase more slowly than new ammonia capacity based on natural gas.

In any case, natural gas needs from existing sources should be adequate to meet ammonia production needs for many years to come. There are still many undeveloped resources of gas mainly in developing countries where the production of ammonia and nitrogen fertilizers will represent the most economic method of exploiting the natural gas. If the price of gas rises significantly in other regions, particularly in the developed countries, there will be more incentive to develop these new resources.

217

Although it may be that biogenetics and other new technologies will play an increasing role in helping to reduce or improve the utilization of chemical fertilizers, this is not likely to take place for many years. Until that time at least fulfilling our increasing need for food will depend mainly on nitrogen fertilizers produced from ammonia based on natural gas.

7

Natural Gas and Natural Gas Liquids in the Chemical Industry

Walter Vergara

The petrochemical sector is in the aftermath of the extensive restructuring that resulted from changes in raw material prices and increased availability, technological improvements, near market saturation for basic petrochemicals in developed countries, and the emergence of nontraditional producers as major partners in world trade. The availability of cheap, abundant gas and natural gas liquids has played an important and decisive role in the development prospects of the industry and is likely to continue to influence the size and location of new producers. These changes have altered the current situation of the industry and its outlook forcing producers, users, and financing institutions to design new strategies, adapt to these changes, and prepare for the future. The sector is at the doorstep of additional changes that have gradually enabled petrochemicals to compete with and outperform traditional materials in the construction, transportation, and machinery hardware sectors as well as serve as substitutes for steel, aluminum, wood, paper, natural fibers and rubbers. The most important trends in the industry and its relationship to gas and gas liquids are briefly summarized below.

Review of the Current Situation of the World Petrochemical Sector

Trends in Feedstock Use

Nothing typifies more the turmoil of change that has recently affected the industry than the rapid changes in feedstock prices and availability. Within the petrochemical industry, natural gas liquids and naphtha have been by far the preferred feedstocks for ethylene synthesis. Naphtha, although still the main feedstock, should continue a long-term declining trend in favor of the new ethane-based ethylene capacity in gas-rich areas and, to a lesser degree, to unbalanced refinery streams ranging from heating oil to refinery gases. Clearly, naphtha prices are critical to the continuation of this trend. In the short term, the shift in feedstocks has been slowed by the drop in naphtha prices from around $300 per metric

219

ton in 1983 to an average of $150 per metric ton in 1989. The softening of the oil market has taken some of the competitive edge away from natural gas as a low-cost alternative feedstock. While prices remain low, naphtha-based producers located near the markets will be able to compete with new plants based on natural gas liquids.

Based on recent World Bank projections of crude prices, naphtha should come back to the $200/metric ton price range by 1995 (Table 7.1). Natural gas, because it cannot be easily traded will continue to vary considerably in price from region to region, ranging from very low in the Middle East to low in Latin America and the Pacific Basin to high in Western Europe where it is associated with premiums for cleanliness, ease of use, and high opportunity values. The value of ethane/propane and other higher molecular weight hydrocarbons typically depends on the cost of extraction and the availability of alternative markets. Overall, natural gas fractions will remain the preferred feedstock for new ethylene plants. Gas-rich developing countries with substantial domestic markets on which to base world-scale petrochemical complexes are poised to gain the most from the trends in feedstock prices. In fact, gas-rich countries have continued to add new gas-based capacity, and new plants under construction or announced in India, Thailand, Venezuela, and Canada are based on natural gas fractions.

Current and Foreseeable
Supply/Demand Situation

The supply/demand situation for basic olefins and methanol (the key chemical derivatives from natural gas) are reviewed in the following paragraphs. The role of gas in the synthesis of these chemicals is reviewed in the following section.

Table 7.1. Current and Projected Prices for Naphtha and Natural Gas (US$, 1988)

	1988[a]	1990	1995
Naphtha ($/Mt)	155	168	188
Natural Gas, Wellhead USA ($/MMBtu)	1.40	1.87[b]	2.19
Factory Gate, Western Europe	2.30		
Middle East, Petrochemicals, Feedstock	0.50	0.50	0.50

[a]Based on World Bank estimates of future crude oil prices. 1 MMBtu = 1,054 MJ.
[b]US Gulf Coast, AGA projections, 1988.

220

Basic Olefins. Ethylene is clearly the main building block of the petrochemical industry and the focus of attention of most countries with petrochemical interests. Further, despite the gradual reduction in its growth rate, ethylene remains dominant in demand and capacity in the world market, as shown by the growth rates and current consumption. Restructuring in ethylene, particularly in Western Europe and the United States, brought substantial reductions in capacity during the mid-1980s. In 1987, the world nameplate capacity was estimated at 46 million metric tons, down from a peak of 52 million metric tons in 1980, while demand was estimated at 44 million (this does not include Eastern Europe, which had an estimated installed capacity of 7 million metric tons). Since then, additional capacity has been brought into stream to meet a demand that grew to an estimated 48 million tons in 1988. Given the softening in crude oil and natural gas prices during the late 1980's, there was speculation on a resurgence in world ethylene growth. But the past shows that oil prices have not had a cause and effect relationship with ethylene demand. On the contrary, the maturation of ethylene-based products has resulted in ethylene nearly reaching the saturation point in the industrialized economies, and this will not change solely on the basis of fluctuations in the price of feedstocks. But the recent surge in polyolefins consumption, coupled with the effects of rationalization and a reluctance in developed countries to invest in new capacity, resulted in a near-record percent of operational capacity for ethylene producers (Figure 7.1) and, consequently, in substantial price increases in 1987 and 1988.

The increase in margins and consequently the higher level of revenues available to manufacturers have attracted a number of potential investors and announcements for expansions and new units by established producers. In the US, for example, from 1985 to the beginning of 1988, not one announcement for new cracking capacity was made, while during the period March 1988-mid 1989, 16 new crackers were announced with a total annual capacity of nearly 6.6 million tons (Table 7.2). This is equivalent to 13% of worldwide installed capacity (excluding Eastern Europe). Although it is doubtful that all the announced capacity will be built in the proposed timetable (some of the announcements may have been done to prevent new entries into the market), the prospect of a large new chunk of capacity entering into operation in the US in the next three to four years will undoubtedly shape the medium-term market conditions for olefins.

A similar surge of new projects has been announced worldwide. In Asia alone, a total of 13 new crackers and expansions adding up to 3 million tons of annual capacity of ethylene are now at various stages of construction and many others are under planning. Announcements for new

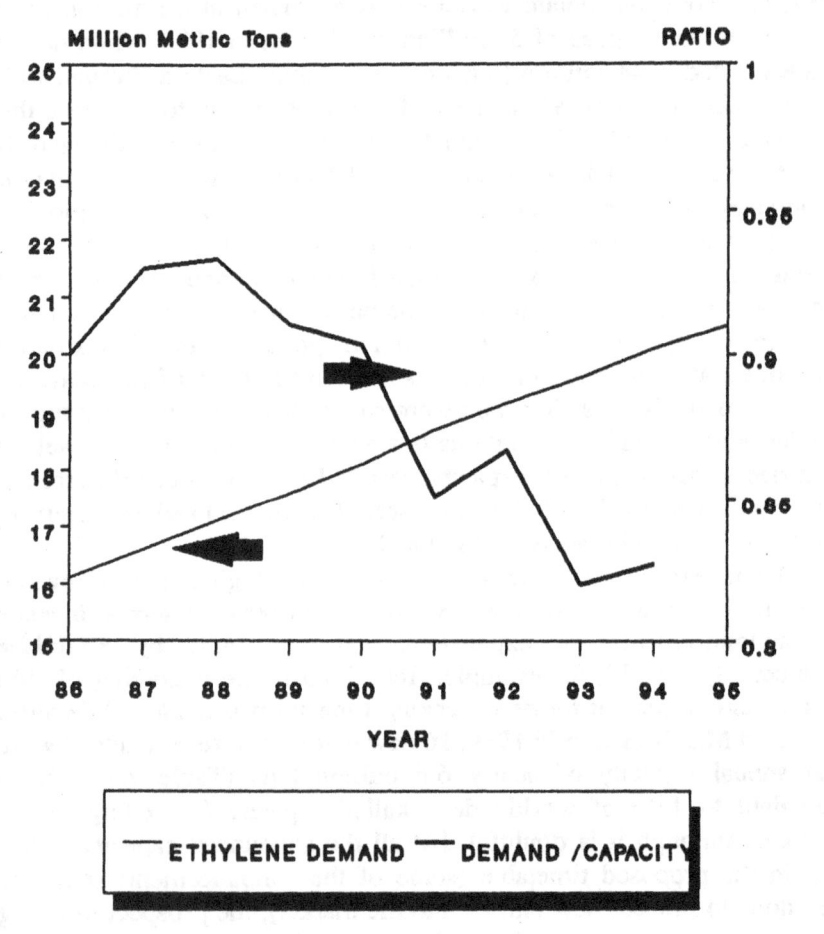

Figure 7.1. Projection of Demand and Production to Capacity Ratio, North America

Million Metric Tons

RATIO

ETHYLENE DEMAND   DEMAND /CAPACITY

Demand projection based on 3.0-2.2%
gradually decreasing growth rate.

Table 7.2. Ethylene Expansion Announcements in the US
(As of Mid-1989)

Target Start-Up Year	Capacity ('000 tpy)	Number of Units
1989	765	4
1990	810	3
1991	1,750	3
1992	1,300	2
1993	650	1
1994	450	1
Other	900	2
Total	6,625	16

Source: Crouch, J., 1989.[1]

ethylene capacity in the Middle East total 1.3 million tpy, in Latin America 2.0 million tpy and in Western Europe 1.5 million tpy. However, these announcements do not match in investment volumes or production capacities the large increases expected from the US and Asia. Overall, this flurry of activity raises the concern that the ratio of production to installed capacity is likely to fall in the next years as can be seen from a plot of announced capacity additions against projected market growth for ethylene in North America (Figure 7.1).

Worldwide, ethylene demand has been projected to grow at an annual rate of 2.9% up to 1995, with developing regions growing at faster rates and Europe, Japan and North America at a much slower rate.[2] A comparison of the expected 1995 demand with the updated numbers for capacity under construction and planned confirms that N. America and the Middle East are likely to have surplus capacity by 1995 (Table 7.3). North America is likely to continue being the price setter in the industry given its surplus situation and the large share of world production capacity. The region accounts for close to 31% of world production of olefins, 34% of commodity plastics and 37% of synthetic rubbers. The data also show that the Asia region as a whole will easily utilize all new capacity under construction and still remain a net importer. Latin America and Africa will remain dependent on imports to meet regional demands.

Methanol. The current world situation for methanol can be characterized as one of plentiful supplies capable of meeting expected demand well into the next decade. During 1986 and 1987, oversupply

Table 7.3. Projected World Demand and Supply Situation for Ethylene by 1995

	Projected 1995 Demand		Projected 1995 Capacity (MMTY)		Ratio of Projected Demand/Projected Supply[b] (%)	
	(MMTY) (A)	Annual Growth Rate 1995/1987 (%)	Current &- in Con- struction (B)	Current, in Con- struction, & Planned (C)	(Operation Factor) (A/B)	(A/C)
North America	19.2	1.9	24.4[a]	24.4	78	78
W. Europe	14.7	1.4	15.5[a]	15.5	95	95
Japan	5.1	1.3	5.4	6.9	94	74
Asia	10.2	8.9	7.6	10.6	134	95
Latin America	5.6	6.8	4.0	5.1	140	109
Africa	1.5	12.4	0.6	1.1	250	136
Middle East	2.7	3.9	4.0[a]	4.0	67	67
Total	59.0		61.5	67.6	97	87

[a]Includes all announcements.
[b]A 95% ratio is the industry standard for full capacity in operation.
Source: Vergara and Brown, 1988,[2] Vergara, 1989,[3] and staff estimates. Eastern Europe not included in the estimates.

resulted in low capacity utilization and prices. A number of plants were shutdown while others experienced production problems at several sites around the world. This led to some price increases during 1988 that quickly yielded a large amount of debottlenecking and restarting of mothballed plants. There is again oversupply in the market, a situation that is expected to continue well into the 1990's.

The major chemical derivatives from methanol such as acetic acid, formaldehyde and solvents are mature products and are not expected to grow at more than 2% per year. The fuel demand has failed to materialize in the face of lower petroleum costs and environmental concerns and it is now very doubtful that methanol use as fuel will become a major outlet in the next ten years. The only exception to date is in New Zealand where methanol is used as a raw material for gasoline synthesis. The other bright spot in the methanol global picture is the demand for MtBe (methyl t-butyl ether) as an octane enhancer in motor fuels. The current world capacity and consumption situation for methanol is summarized in Table 7.4. Future growth is projected at 3.5% per year. World demand for MtBE grew at 16% per year for the last 2 years. The exceptionally high growth in demand for MtBE is a direct reflection of the fast pace of lead phase-down in the US and Europe. Future high demand is expected to continue as more

Table 7.4. Methanol: World Capacity, Consumption and Future Demand (Million Tons/Year)

	1987	1990	1995
Capacity	20.4	22.0	23.2
Total demand	16.4	18.2	20.8
Surplus	4.0	3.8	2.4

Source: Crocco, J., 1989,[4] Crocco, J., 1989,[5] and author estimates.

countries apply lead-limiting regulations to motor fuels, and the existing programs target more stringent specifications.

World MtBE output in 1987 is estimated at 4.7 million tons. The combined capacity of planned units will easily meet future requirements up to 1990, but new installations may need to be planned for the early 1990s in order to meet the expectations for new market development. These will largely depend on how fast new lead phase-down programs are implemented in W. Europe and developing countries. For methanol itself no new capacity requirements are anticipated until the mid-1990s.

International Trade

The growth rate in world chemical trade has been as extraordinary as the growth in chemical production. World trade of chemicals grew from $22 billion in 1970 to $216 billion in 1987, and of this an estimated 60% (or about $130 billion) is for petrochemicals. In 1987, chemical exports from Western Europe and North America amounted to $130 billion and $31 billion, respectively. International trade in chemicals continues to be dominated by industrial nations, although new producers from developing countries are beginning to have a significant impact in trade statistics. The principal partners in chemicals trade are summarized in Table 7.5 along with the associated volumes of exports and imports.

As a major consequence of the oil price increases, gas- and oil-rich developing nations saw an improvement in the competitiveness of their petrochemical trade. Between 1970 and 1985, the developing countries' share of chemicals exports increased from 4.6% to 6.0%. New producers, notably from Asia, affected world trade in two different ways: (a) products from the new petrochemical manufacturers replaced previously imported products, which had been supplied by industrialized countries, in their domestic markets; (b) the new producers entered into the world market and competed for market share in developed countries with some success. Still, as Table 7.5 shows, developing countries are far from being a dominant

Table 7.5. Regional Composition of World Trade in Chemicals,[a]
(Current US$ Billion)

	1980		1987	
	Exports	Imports	Exports	Imports
North America	26.6	11.6	31.0	21.5
Western Europe	92.0	73.2	129.6	101.9
Japan	6.7	5.9	11.6	11.8
Eastern Europe & USSR	7.8	10.6	11.1	15.0
Developing Economies	9.8	38.2	13.1[b]	39.8[b]
Others	5.1	8.5		

[a]World trade was $148.0 billion and $215.7 billion in 1980 and 1987, respectively.
[b]For 1985, the last year for which statistics are available.

force in the world market of chemical exports and, despite the recent changes, are still one of the largest outlets for production surpluses from the industrialized nations.

Trends in Production Technology

Technological progress has been a key factor in the remarkable resilience of the petrochemical industry to changes in the conditions of the world market. For example, the sharp increases in feedstock costs of the 1970s were gradually but effectively neutralized by corresponding increases in total product yields and reductions in specific energy consumption in cracking operations (Figure 7.2). There have also been major changes in downstream processes, such as the successful commercialization of gas-phase polymerization processes and the automation and diversification of extrusion technology and the development of composites and blends. All of these changes have contributed to a more efficient and productive industry that is better capable of withstanding cyclical price changes. Likewise, the gradual maturity of a number of downstream petrochemicals that resulted in a great deal of concern for the future growth of the industry has sparked a new generation of products. New plastics, rubbers, and fibers, with substantial advantages in production costs, improved performance, and technical characteristics, have all appeared in the world market.

The new products reaching the market in industrialized nations can potentially substitute for long-standing traditional materials in many sectors of economic activity. Therefore, the lower per capita demand for some traditional materials (like steel or natural rubber) in developing

Figure 7.2. Evolution of Yields at Ethylene Crackers

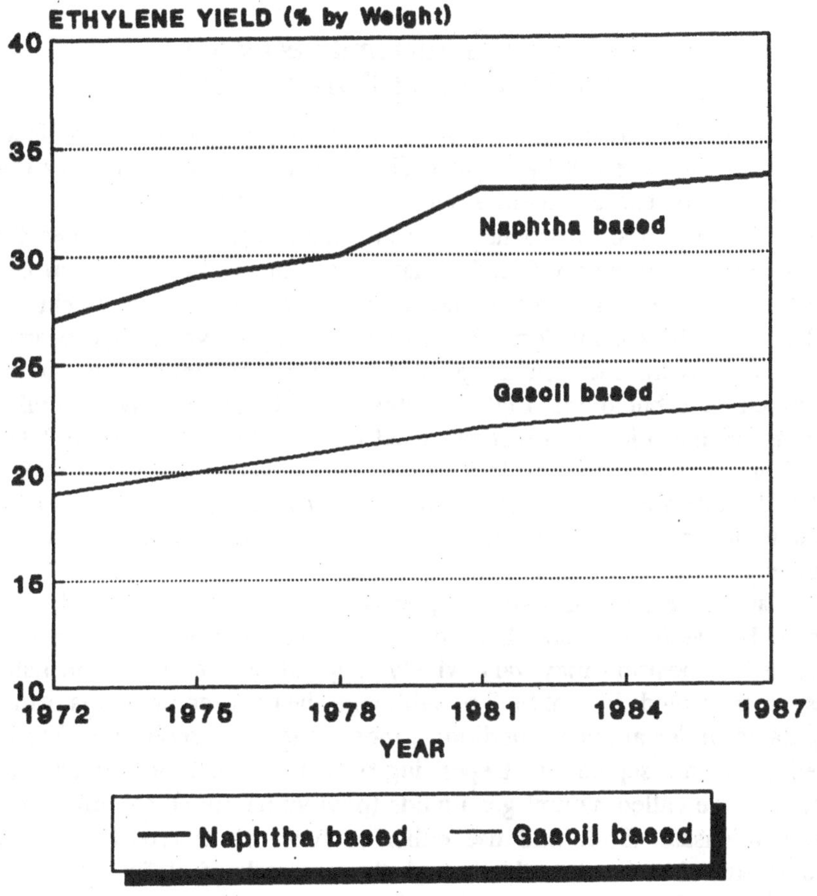
ETHYLENE YIELD (% by Weight)
40
35
30
25
20
15
10
Naphtha based
Gasoil based
1972
1975
1978
1981
1984
1987
YEAR
Naphtha based
Gasoil based

countries offers an opportunity and challenge to those economies to capitalize in technological improvements at an earlier stage of their development. Much of the current activity in technological development is also related to the introduction of more versatile technologies and processes capable of producing a wider variety of polymer grades. Other current technological improvement priorities are increases (a) in catalyst activity and life span, (b) in use of common co-monomers such as butene to develop additional products, and (c) in process retrofits.

Gas Role as Fuel and Feedstock in the Synthesis of Petrochemicals

Worldwide, the chemical industry is the largest industrial application of natural gas, where it has historically played two complementary roles: (a) a source of energy (thermal and mechanical power) and (b) as a raw material. Most basic petrochemicals are manufactured or obtained from hydrocarbons commonly found in natural gas or cracked from petroleum in refinery operations. A major class of products of interest to the chemical industry are the olefins derived from low-molecular-weight hydrocarbons through cracking and dehydrogenation. The major olefins are ethylene, propylene and butadiene. Light naphtha, gas oil fractions and natural gas liquids are the primary feedstocks used in their synthesis (Figure 7.3). In refineries a second major class of petrochemicals produced through the catalytic reforming of naphtha is aromatics. These include benzene, toluene and xylenes and comprise about half of the reformed naphtha or reformate.

Natural gas, as discussed in previous chapters, is composed of low molecular weight saturated hydrocarbons. Depending on the source of the gas, its composition may vary widely. Typical ethane content in natural gas is presented in Table 7.6. Although heavier components can be separated under ambient conditions, most of the low molecular fractions need cryogenic separation. Depending on the composition, the separated fractions are called natural gas liquids (predominantly ethane/propane) or gas condensate. Both find use either as fuels or as feedstock for the petrochemical sector, mostly through the synthesis of olefins. The major component of natural gas, methane, finds use also as fuel and as feedstock for chemicals, either in the synthesis of ammonia (already discussed in Chapter 6 and therefore not presented in this analysis), in the production of methanol and derivatives, and in the synthesis of a wider range of other chemicals such as acetylene, halogenated methanes, phosgene, and others. In the section below, the role of gas as feedstock for synthesis of each of these products is briefly reviewed.

228

Figure 7.3. Feedstocks Used in the Synthesis of Ethylene
(as Percent of Total)

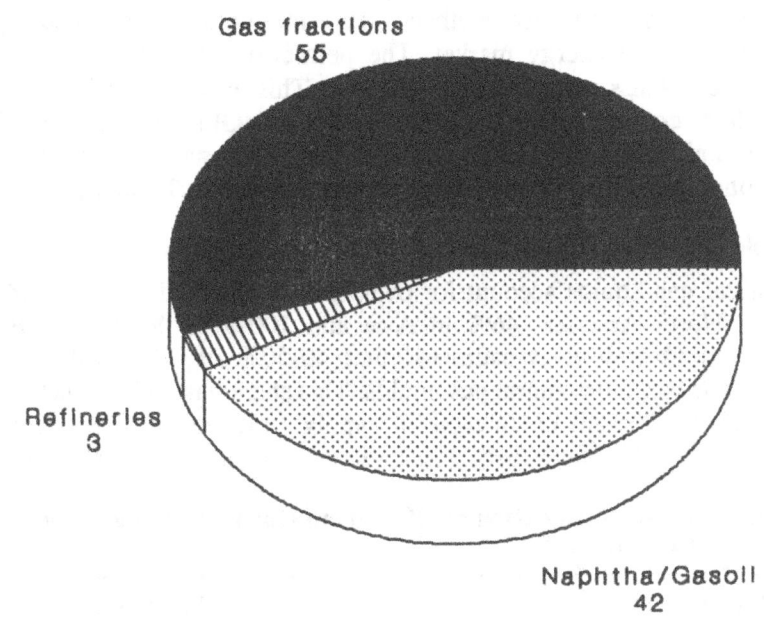

Gas fractions
55

Refineries
3

Naphtha/Gasoil
42

**Table 7.6. Typical Ethane Fraction in Natural Gas
(% of Total Volume)**

	%
Nonassociated gas	0.5-2.5
Associated gas	below 7.5
Rich gas	above 7.5

The sources of natural gas liquids are both gas processing plants and refineries, although the vast majority of NGLs is now obtained from gas fields (close to 95% in 1986). Its supply is therefore associated to the production of gas (mostly methane) and the demand for other gas derivatives for the energy market. The petrochemical sector thus must compete for NGLs with alternative users. This is particularly true in energy-short countries with multiple uses for NGLs in the domestic, fertilizer and refinery sectors. The estimated regional production and consumption of NGLs in 1986 are summarized in Tables 7.7 and 7.8.

Gas Role in Olefins Production

Natural gas liquids account for over 68% of the feedstock to olefins plants in North America, 46% in Asia and 100% in the Middle East (Table 7.9). Overall it is estimated that 40-50% of all world olefins production is based on natural gas liquids. This share has widely fluctuated in the past as a result of changes in supply, competition from heavier feedstocks and alternative uses. Still, ethane remains the most efficient

**Table 7.7. World Production of NGL from Gas Processing Plants,
1986 (MMT)**

Region	Production
North America	58.8
Western Europe	8.8
Developing Asia	4.8
Latin America	15.8
Middle East	20.1
Other (Eastern Europe and Africa)	35.9

Table 7.8. World Production of NGL at Refineries by Region, 1986 (MMT)

Region	Production
North America	15.0
Western Europe	13.6
Japan	4.2
Developing Asia	4.2
Latin America	5.5
Middle East	2.5

feedstock for ethylene and natural gas liquids remain the preferred olefins feedstock in regions where low-cost production of this material is available. The estimated regional production of olefins and the share of this production that is gas-derived are summarized in Table 7.9.

Steam cracking of low molecular weight hydrocarbons is the preferred process for manufacture of olefins. Ethylene is also recovered from refinery gases and a few plants obtain ethylene from ethanol dehydration (in India and Brazil) and from coal (S. Africa). These last two routes are not competitive with steam cracking.

With the changes in oil prices and its effect on naphtha prices, new producers are increasingly opting for wide flexibility in feedstock utilization in the design of olefins plants. Nevertheless, gas-rich or gas-surplus areas with established gas-gathering and processing capability are expected to continue to favor the use of natural gas liquids in the synthesis of olefins. Countries where gas availability is expected to continue to favor gas use in petrochemicals include Canada, gas-rich

Table 7.9. Production of Ethylene by Feedstock Source, 1988 (%)

Region	Total Production (MMT)	Gas Liquids (%)	Naphtha & Gas Oil (%)	Refinery Gases (%)
N. America	20	68	29	3
W. Europe	15	7	91	2
Japan	5		100	
Asia	2	46	52	2

countries in Asia and the Middle East. When combined with low capital costs these countries are expected to remain or become part of the most competitive producers of olefins in a market that continues to grow at high rates (Table 7.10).

Capital costs for steam crackers using gas fractions are considerably cheaper than corresponding costs for plants cracking heavier feedstocks. This is mainly because the use of ethane/propane precludes the need for separation and recovery equipment for all the by-products and offstreams generated by the use of heavier feedstocks. Lower costs contribute, given the capital intensive nature of the industry, to significant savings in manufacturing costs. The estimated cost structure of ethylene manufacturing from gas liquids and naphtha is briefly summarized in Table 7.11 for US-based producers.

Gas Role in the Production of Methanol

Production of methanol is the second largest use of gas in the chemical sector. It is commercially produced by reacting synthesis gas over a catalyst. Synthesis gas in turn is obtained primarily from reforming of natural gas, but also from other hydrocarbon sources such as naphtha and heavy fuel oil. Nonetheless, the most competitive route is through the use of natural gas. Natural gas-based processes require about 1.0-1.5 m^3 of gas per kg of methanol produced. Synthesis gas can also be produced from coal liquefaction but relatively low prices for gas and oil have prevented wide commercialization of coal-based processes. Only a few coal-based plants are in operation, mostly in South Africa. The estimated cost structure of methanol manufacture from natural gas is briefly summarized in Table 7.13.

Table 7.10. Current and Projected World Demand for Olefins (MMT)[a]

	1988	1995
Ethylene	48	58
Propylene	24[a]	32
Butadiene	6[a]	7

[a]Does not include Eastern Europe.

232

Table 7.11. Estimated Ethylene Production in the US Gulf (US$/Ton)

	Naphtha-Based	Gas-Based
Raw materials	216	90
Labor	5	4
Overhead	1	9
Other	16	43
Total	247	146
Depreciation	111	105
Interest	55	56
Return on equity	98	100
Total	540	430

Based on a 600,000-tpy ethylene cracker, linear depreciation at 10 years and 2.5 years of construction. Includes by-product credits.

Gas Role in the Synthesis of Other Chemicals

Acetylene manufacture consumes about 1-2% of natural gas used in the chemical sector in the US and is estimated to require a similar fraction of gas used in industry in Western Europe. The most widely used production process for acetylene is the oxidation of natural gas. Demand for acetylene by the chemical industry has significantly been reduced from the levels of a decade ago as a result of increased reliance on ethylene. On the other hand, natural gas has now replaced in most countries the use of

Table 7.12. Projected Demand for Methanol and Gas Requirements By Region, 1995 (MMT)

	Methanol 1995	Natural Gas Requirements 1995
North America	6.4	2.9
Western Europe	6.1	2.8
Japan	1.8	0.8
Developing Asia	3.1	1.5
Latin America	1.0	0.5

Table 7.13. Estimated Production Cost for Methanol (US$/Ton)

	Cost
Raw materials	40
Labor	5
Utilities	18
Overhead	26
Maintenance and Insurance	10
Depreciation	35
Interest	25
Return on equity	36
Total	**195**

US Gulf conditions for a 600,000-tpy facility, linear depreciation at 10 years and 2.5 years of construction.

---

calcium carbide as raw material for acetylene and the switch has made acetylene production more competitive. Little change in total requirements of gas though is expected in the near future.

Oxochemicals include aldehydes, alcohols and other chemicals in the C3-C12 range. The most commonly used are plasticizer alcohols, aldehydes used in the manufacture of detergents, n-butanol and others. Methane is the main source for synthesis gas used in the manufacture of oxochemicals. Oxochemicals account for over 0.5% of all gas used in industry. The largest fraction of gas used for oxochemicals is consumed for the production of butyraldehyde (over half of the world production of oxochemicals).

Chlorinated methanes account for about 0.1% of natural gas used in the US chemical industry. These are used in the manufacture of fluorocarbons, tetramethyl lead and for direct use of chloroform and carbon tetrachloride. These materials are mostly produced via thermal chlorination of methane. Controversy regarding the impact of fluorocarbons in the ozone layer and evidence regarding the impact of leaded compounds in human health have led to continuous reductions in the demand for these materials. The United Nations, Western European governments, Canada and the US have all taken steps to reduce the production and emission of fluorocarbons and leaded compounds. Further reduction in the demand for these materials is expected pending enactment of new emission performance standards.

Other chemicals derived from natural gas include hydrogen cyanide mainly used as raw material for adiponitrile and carbon disulfide which is used in the production of cellulosic fibers and pesticides. But, the world consumption of natural gas for these materials is very marginal. The estimated uses for natural gas in the chemical industry are also shown in Figure 7.4; the share of natural gas uses in the chemical sector is summarized in Figure 7.5.

Economic Implications

There are many factors influencing the use of natural gas in the synthesis of petrochemicals and the outlook for its continuing demand. These include feedstock price and availability, capital costs and prospects for market growth.

Feedstock prices are a key factor to economic competitiveness of petrochemical production. For purposes of the analysis, future naphtha prices consistent with the World Bank projections of future crude costs are considered. These assume that oil will increase from US$14 per barrel in 1988 to about US$22 per barrel in 2000 (in constant 1988 terms). Assuming that refinery margins remain constant at levels which ensure adequate returns to refiners, prices in naphtha in the spot market are expected to increase from about US$140 in 1988 to about US$200 by 2000. It is possible, however, that increased demand for naphtha from the petrochemical industry and the impact of lead phase-downs in gasoline expected in a number of countries may result in higher refiner margins and may drive up naphtha prices.

Because gas is difficult to transport and not easily tradeable, the valuation of ethane is very site-specific. For gas prices, valuation is based on the opportunity cost of gas and its fractions in each country. In locations with large gas surpluses, the long-run marginal cost is normally the determinant of gas valuation.

Estimated ethane economic price in some countries in Asia, recently estimated by following the procedure outlined above are summarized in Table 7.14. These include countries with gas surpluses (Malaysia, Indonesia) and some where gas although available has alternative uses.

Feedstock Availability

While ethane remains the preferred feedstock when available, the olefins industry has been increasingly building cracking capacity with flexibility to use a wide range of feedstocks to take advantage of international market shifts in availability and relative prices. Still, many of the largest petrochemical producers also have substantial volumes of gas

235

Figure 7.4   Chemicals from Natural Gas

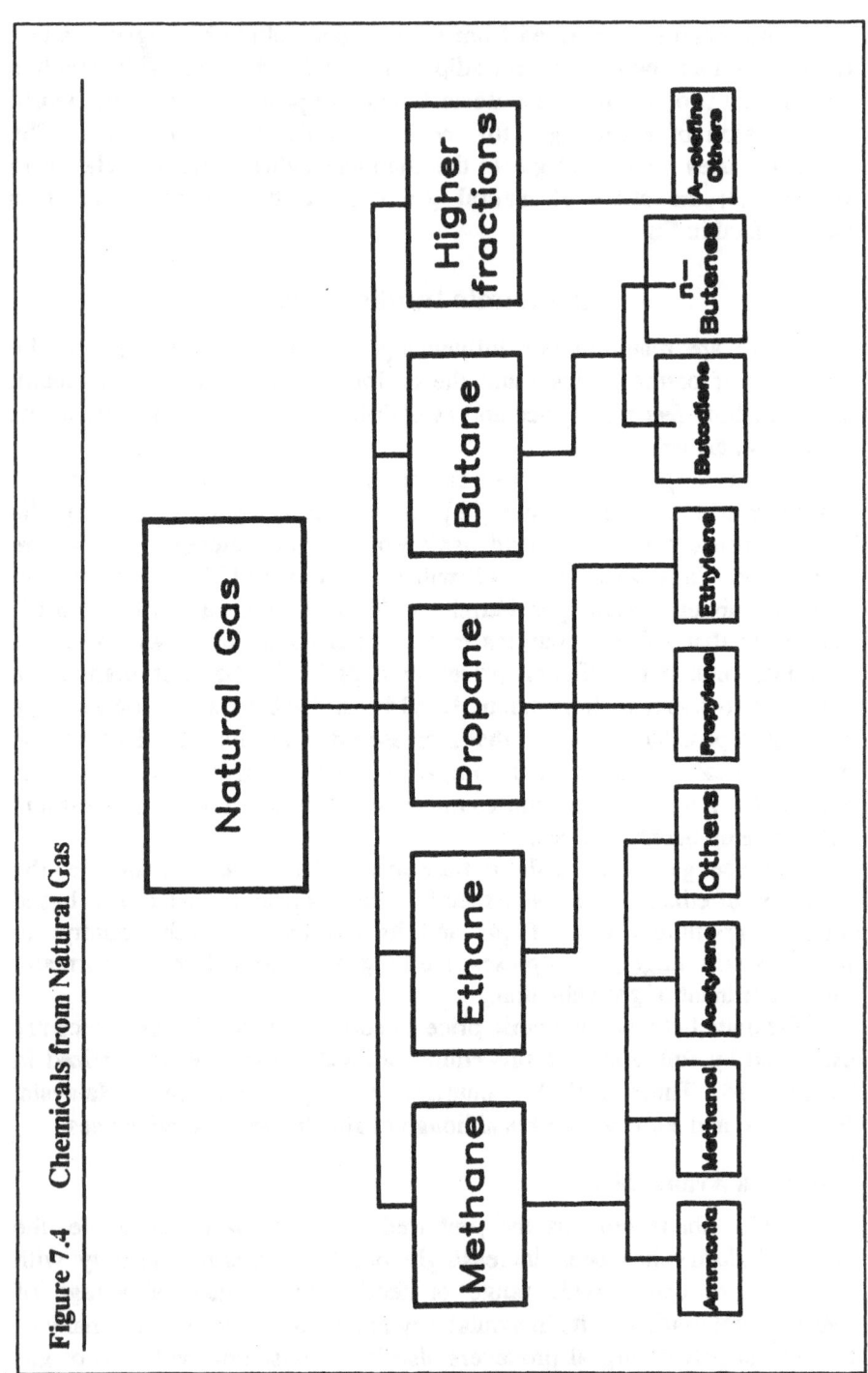

Figure 7.5 Gas Based Chemical Production Share of Gas Use

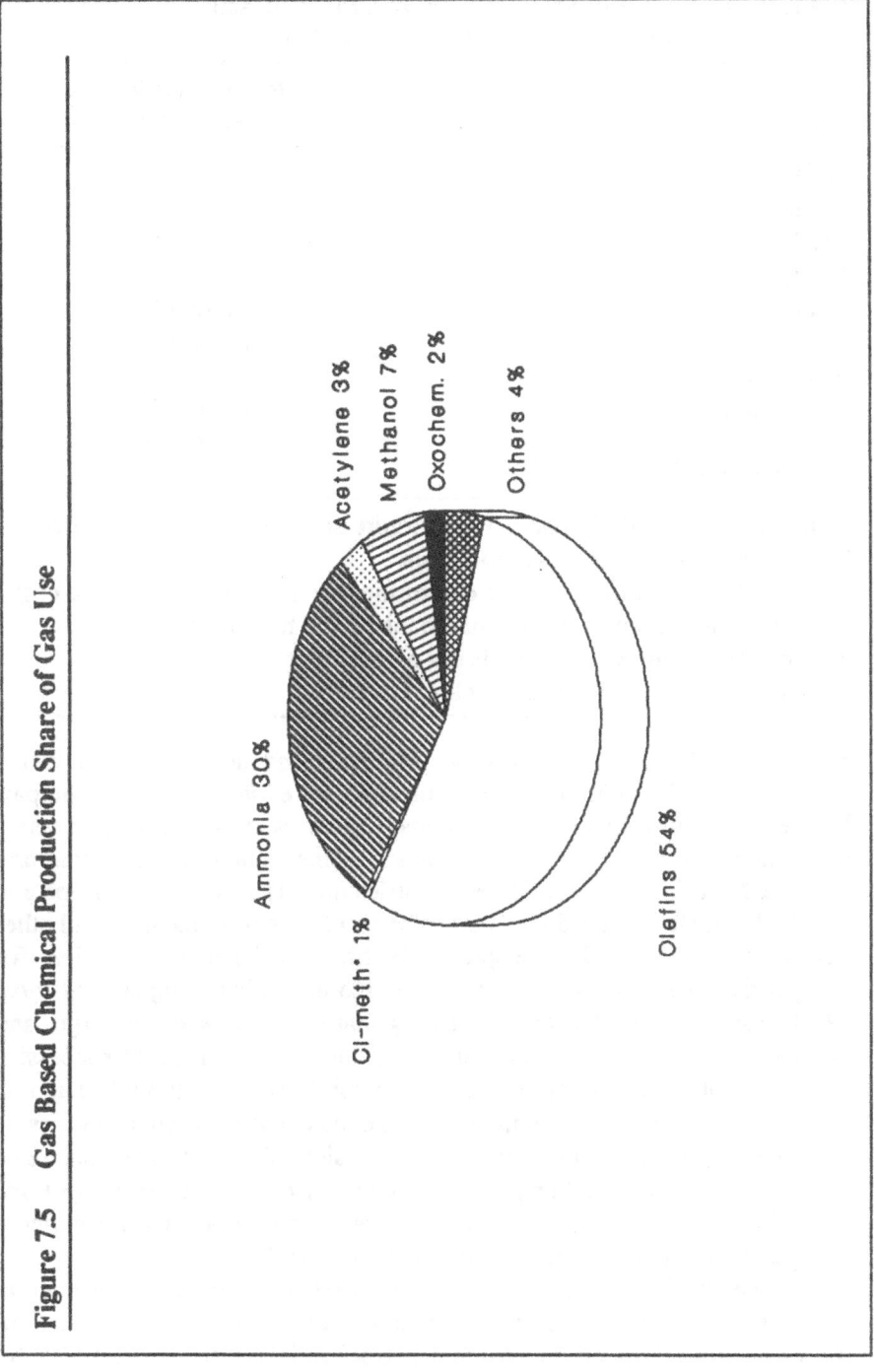

Ammonia 30%

Cl-meth. 1%

Acetylene 3%

Methanol 7%

Oxochem. 2%

Others 4%

Olefins 54%

Table 7.14. Projected Ethane Economic Price in Asia (By Country, in 1988 US$/MMBTU)

Country	Projected Price 1988 to 2000
India	3.0 to 4.5
China	3.0 to 4.5
Thailand	4.0 to 5.5
Malaysia	
Peninsula	2.5 to 3.5
Sarawak	2.0 to 2.5
Indonesia	
Sumatra	2.0 to 2.5
West Java	2.0 to 3.0
East Kalimantan	2.0 to 2.5

India, China, and Thailand: Opportunity costs estimated at imported fuel oil equivalent plus separation costs.

Malaysia: Opportunity value determined by its long-run marginal cost.

Indonesia: Opportunity cost of ethane linked to production and distribution of gas since supply is not a constraint.

Source: Vergara, W. and Babelon, D., 1989.[6]

resources (Table 7.15). For example, Malaysia, Indonesia, India and China in Asia have all ambitious petrochemical sector expansions, driven in part by the rich endowment of gas fractions. Countries in other regions present the same circumstances (Canada, Mexico, Venezuela in the Americas and Algeria, Nigeria in Africa). Other countries notably in Western Europe and the Far East, already produce 100% of their olefins from naphtha and other petroleum fractions. In the past this has not been a limitation for competitive manufacture. For example, Korea, Taiwan, Japan and most West European producers all lacking natural gas, keep a large and competitive industry, whose products are used as key inputs to successful export-oriented manufacturing sector. In the long term, though, naphtha availability for countries with no indigenous hydrocarbon resources is expected to play a limiting role in the expansion of the industry because of tighter supplies and higher prices. This seems to be particularly true for countries such as Korea and Taiwan where the future demand for naphtha is expected to increase significantly over the next decade.

Lower feedstock costs in gas rich countries translate into a substantial advantage in production costs of ethylene and significant cost savings in the production costs of downstream petrochemicals (this is after all why

238

Table 7.15. Oil and Gas Production and Reserves in Gas-Rich Developing Countries in Asia, 1988

| | Production | | Reserves | | |
	Oil ('000 Bbl/Day)	Gas (MCFD)	Oil (Mbbl)	Gas (BCF)	Number of Refineries
China	2,682.0	1,420	23,550.0	31,700	40
India	609.0	1,000	6,354.2	22,861	12
Indonesia	1,186.0	1,740	8,250.0	83,590	6
Malaysia	2,309.0		2,922.0	51,700	4
Thailand	31.9	490	85.2	3,900	3
Brunei	139.0		1,400.0	11,600	1

Mbbl: Millions of barrels.
BCF: Billions of cubic feet.
Source: *Oil and Gas Journal*, 1988,[7] and staff estimates.

Canada and Saudi Arabia have become world-scale producers of downstream products in recent years). The expected increases in naphtha and natural gas will improve the margins available to gas-rich or naphtha-surplus countries or to countries with low long-run marginal costs. In 1990, at currently forecast prices, gas-rich countries with access to natural gas at US$1.00/million Btu or less (such as Indonesia, Algeria, Nigeria, Venezuela, Argentina and others) will be able to produce basic olefins with a cost advantage if the infrastructure, management and financial constraints are satisfactorily addressed. For instance, ethylene production costs may be about 30% lower in these gas-rich countries compared to similarly sized and efficient producers in Western Europe or naphtha importing countries such as Japan and South Korea. Basic commodity polymers could likewise be produced with cost savings in the 5-15% range.

The gradual switch to lighter feedstocks, if naphtha prices increase in the future, will impact the availability of propylene and butadiene. Yet, the use of proven processes for the recovery or synthesis of propylene from alternative sources such as refinery operations and propane imply that natural gas liquids as a feedstock for petrochemicals will no longer limit the production of propylene. If future prices of propylene increase as a result of feedstock limitations, then enough economic incentives will be at hand to allow for propylene production from a variety of sources. Further into the future, and depending on the economics, crude oil cracking may also affect the relative supplies of ethylene and propylene by enabling the use of heavy fractions as feedstocks for cracking. Liquefied petroleum gas availability is limited by the level of oil production and the price of naphtha. In volume terms its use as a petrochemical feedstock is also expected to decrease relative to the use of natural gas.

Aromatics, on the other hand, are produced by extraction from oil refinery streams resulting either from the processing of crudes with high aromatic content or from the cracking and reforming of refinery fractions. While the precise operating conditions can be tuned for optimal production of benzene, the overall economics of the cracking or reforming process and therefore the availability of aromatics feedstocks are usually determined by the volume and octane quality requirements of gasoline. The widespread lead phase-down programs for automotive fuels in the industrialized countries have placed a premium on the value of these feedstocks as octane enhancers for gasoline and increased the forced a competition for aromatics between the fuel and chemical industries. Aromatics are also tied to the production of synthetic fibers, xylenes and rubbers.

Geographical Location

Geographical location can only be counted as an advantage if it translates into competitive access to markets or raw materials. Asia and Latin America are expected to continue to be as a region a net importer of a wide spectrum of petrochemical products. Sizeable import markets are expected to continue in China, Indonesia and India; while prospects for increased share of imports in the domestic markets are expected to continue in China, Indonesia, India and the Andean countries, prospects for increased share of imports in the domestic markets of Japan, and Taiwan are also expected. If tariff barriers and other import restrictions are discounted, freight charges emerge as an important element in delivered costs. A spot check of current freight charges shows that established producers in Asia have a significant saving advantage (as much as 5-10% of FOB price) against manufacturers in the US and Canada. It is also expected that future Japanese and Taiwanese manufacturers will continue to rely on competitive imports from subsidiaries or other companies located in less developed countries in the region.

Domestic Pricing Policy

In a number of developing countries, prices of petrochemicals are significantly higher than international prices. These higher prices are the result of several, interlinked policy decisions and the pattern of development. The petrochemical industry is capital- and energy-intensive, typically requiring large capacities to gain economies of scale.

240

Petrochemical products are primarily used in relatively sophisticated applications that require a somewhat more developed economy. Thus, domestic production is more likely to be interesting to the larger developing country with an already well-developed industrial base.

The most typical approach that has been employed in developing countries is to give the sector an infant industry status and to develop the industry with a relatively high level of protection. Quite often there is a parallel capital goods/steel industry that also enjoys a high protective trade regime. To minimize initial capital outlays there is the temptation to build smaller, less than minimum economic sizes and/or to locate smaller (therefore more) units in various parts of a country to satisfy political/social pressure for regional industrial/economic development. These two factors add to high capital cost due to import taxes on equipment plus domestic capital goods inefficiencies and the diseconomies of smaller capacities.

Each of these factors contribute toward higher costs, higher import duties and higher price for petrochemicals. This result can be an industry that is not as competitive as desirable. Several developing countries, e.g., Brazil, China Indonesia and India, have begun to break away from such restrictive policies while others such as Korea have already a relatively open and competitive industry, but the transformation to a fully competitive environment will typically be a lengthy process, given the complexity of already established policies and practices. However, each country has a different policy and industrial environment that requires a country-specific analysis before any firm conclusions can be drawn.

Long-Term Potential of Gas in the Chemical Industry

The future use of gas in the chemical industry is conditioned by four important factors:

- The availability of the appropriate quantities of natural gas fractions, and economics of separation plants;

- The relative pricing of gas and gas fractions relative to the price of alternative feedstocks such as naphtha, coal and gas oil;

- The long-term development prospects of the chemical industry, and the limitations imposed on the use of chemical and feedstocks due to environmental conditions.

The issue of gas availability has already been discussed in the previous section. Overall, it is not an issue in the short- and medium-term development of the industry. Considerable gas reserves and substantial improvements in gas gathering, processing and transportation infrastructure is available among current and potential petrochemical producers. In this section the other factors, namely, relative pricing, development prospects and environmental considerations will be discussed.

Relative Feedstock Pricing

The price level at which olefin producers would switch from ethane/propane to heavier feedstocks depends of course not only on the relative prices of the feedstocks, but also of the co-products yields and prices, capital and energy costs and installation factors.

The crossover price or the price at which a new manufacturer is indifferent to either feedstock assuming all other factors constant has been estimated for a US-based producer and is shown in Figure 7.6. Clearly, the crossover price is rather theoretical as each location has different circumstances and is cost-specific. The concept is also applicable only to new units or to producers that have already built-in capacity to handle different feedstocks. Nevertheless, the results can be used as illustrative of the advantages provided by feedstock pricing. The results show that under the current naphtha market conditions domestic manufacturers in Asia can afford prices for ethane of up to US$3.7/MMBtu and that this range will increase to US$4.8/MMBtu if naphtha price projections materialize.

One primary consideration in the determination of advantages of low cost gas based producers is the extent to which freight costs can be supplemented, absorbed, by differences in feedstock costs or capital-related costs. Table 7.16 summarizes differences in the key cost elements that can offset freight costs for polyethylene, an ethylene derivative. The data illustrate the relative importance of different factors in the establishment of advantages by low cost gas based producers in Asia. Large variations in feedstock costs (500%) and smaller but important differences in installation factors (30-35%) and capacity utilization (15%) can be compensated by a US$100 freight difference (the estimated freight cost from the US). The analysis assumes that Asia remains a net importer of olefin-based petrochemicals the chief competitor of Asian producers remains the United States.

In summary, for net import countries building olefin capacity for their domestic markets, a US$100 freight advantage shifts the available cost of feedstock (that is, the cost of ethane at which domestic production would still compete with landed cost of US imports), up by about US$2.0 per million BTU for ethane, but only if everything else is equal, in particular, cracker sizes, capital costs and capacity utilization similar to the US. Once

Figure 7.6. Crossover Price for Gas and Naptha as Feedstock for Olefin Synthesis

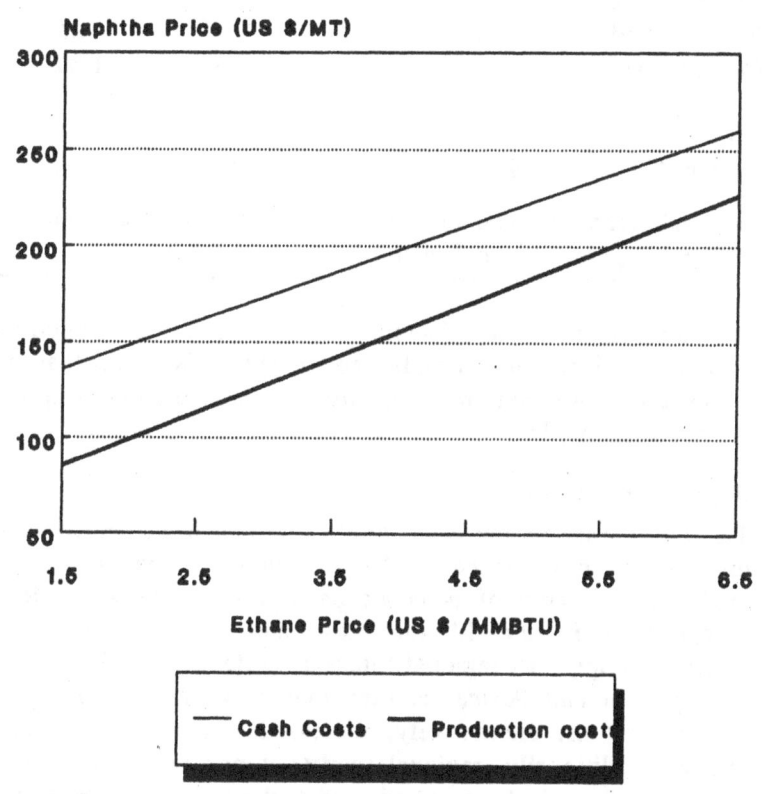

(USA Gulf based location

with Propylene at US $ 430/MT)

Prod. Costs include a 20% ROI

Table 7.16. Differences in Feedstock Costs and Capacity Utilization That Compensate for Freight Advantages for Polyethylene

	Freight (Per Ton of Polyethylene)				
	US$20	US$25	US$50	US$75	US$100
Ethane (US$/mmbtu)	0.4	0.7	1.0	1.6	2.0
Installation Factor					
Ethane-Based	1.10	1.15	1.20	1.25	1.35
Naphtha-Based	1.05	1.10	1.15	1.20	1.30
Capacity Utilization (%)	96	93	90	86	82

Note: Ethylene cost accounts for close to 70% of polyethylene costs.
Source: Vergara, W., and Babelon, D.[6]

the particular conditions are factored in low capital cost countries with large plant capacities enjoy a higher advantage. This means that crackers for domestic consumption are competitive for ethane costs of up to about US$4.5 per million BTU.

Development Prospects

The world market for basic olefins as has been shown is slowly shifting toward an increased production share by energy rich nations. Although a large fraction of the new capacity installed in the last few years in concentrated in the Middle East and Canada, recent market developments imply that regional foreseeable demands in Latin America, developing Asia and Africa requires major investments in new basic olefins capacity or, alternatively, increased imports of petrochemical derivatives. Additionally, regional market trends encourage developing countries in Asia with the required feedstocks to step up their efforts to become important petrochemical producers for their regional market.

In Asia in particular the prospects for future development of the industry are generally favorable. First of all because the domestic markets are expected to continue to grow and expand soon converting the region in to a market of comparable size to other developed regions. Another reason is the differentiation of markets within countries in the region that provide opportunities for mutual complementation. Finally the abundant gas resources in some countries in the region, and the availability of other feedstocks, provide Asian producers with the possibility to become long term low cost producers of basic petrochemicals and a competitive manufacturer of downstream products. As a caveat, these generally favorable prospects hinge on the continuation of moderate growth on energy prices. Also, the outlook would be affected if additional large volumes of export oriented capacity were to be introduced by countries known to own vast resources of associated gas (such as Iran, Qatar, Algeria and others).

Environmental Considerations

Globally, the sector has proven resilient against a background of changing economic and environmental circumstances. Petrochemicals continue to gain ground in all sectors of the economy and are central to industrial strategies in developed countries. The prognosis is for the industry to maintain a healthy pace of growth notwithstanding periodic cycles.

The widespread adoption of emission standards, the availability of new technologies that minimize effluents and maximize resource recovery and the realization by technology and engineering companies that environmental liability does not cease at plant start up even if local legislation is lacking or lax, has contributed to tight the performance of the industry. But, the steady improvements in point source emissions have been offset by (a) the issue of solid waste generation associated with the end products, (b) the rising concern that manufacturers should share in the responsibility for managing the life cycle for all petrochemical products, and accept liability for the generation of hazardous waste, and (c) the realization that the long-term environmental effects of what were until very recently accepted industrial practices are associated with much higher social costs than was previously considered.

This is evident in the concerns related to the degradability of plastics and its contribution to total solid waste, which have been apparent in recently proposed legislation banning the use of plastic products in some localities in the US and Western Europe. Although the share of plastic products in total solid waste production in the US is below 7% of the total by weight, plastic residue is highly visible and the total volumes involved are considerable. Plastics are in particular vulnerable to criticism. According to a recent estimate, about 40% of total plastics production in the US, ends up in disposables. This would equal about 10 million tons per year, equivalent to the aggregated yearly production of plastics in the whole of Asia.

Recycling is one alternative to waste disposal, which has been used by other industries for a number of years, but presents practical complications when applied to plastics, since these products are subject to stringent specifications which make recycling unfeasible in many instances. Another alternative is the use of recycled materials in new applications. Currently, very little plastic material is recycled in the US (less than a fraction of 1% against 40% of all aluminum and 30% of all paper), but as pressure mounts to reduce the volumes of solid waste associated to plastics, recycling and reuse of plastic materials should increase. The introduction of recycled material even at modest levels will have an impact on the future of the industry by slowing down the rates of growth in consumption of commodity plastics and rubbers.

Incineration is another alternative to plastic disposal. Incineration not only reduces the large tonnage of solid waste (typically 1,000 tons yield 250 tons of ash) but can also generate electricity. There are nevertheless questions yet to be answered that cast doubts about the applicability of incineration to plastics. These have to do with the potential generation of dioxins and furans, the presence of heavy metals from plastic additives and the generation of acid gases from the incineration of PVC. There are also some concerns about the economic viability of incineration with energy prices at the current levels. Incineration is not likely to become a major disposal method for plastics in the short term.

Although a lot of attention is being given to solid waste generation from plastic disposal, the industry is also associated with liquid waste effluents and the emission of airborne pollutants. Comprehensive legislation to deal with these effluents has recently been enacted in the US and Japan. Application of emission and performance standards is nevertheless complicated by the large number of products involved and also by the continuous modification in processes design. Consider, for example, the recently completed source performance standards for the polymer industry in the US, which were released at the time when major innovations in reactor technology (slurry and tubular reactors) were entering commercial application. As a result, the standards are in some cases already not applicable or obsolete.

The main difference between air emissions and water pollution from the petrochemical industry and other manufacturing processes is the heterogeneous nature of the contaminants, mostly of hydrocarbon nature. Many of the compounds are toxic or hazardous and require special handling and disposal. Over the years, a number of techniques have been introduced to reduce hydrocarbon emissions at the source.[8] These techniques have proven useful in the reduction of product loss and resource recovery. The adoption of natural gas liquids as feedstocks has contributed to reduce byproducts and the diversity of emissions. The costs of air and water pollution control systems varies widely with the process and control required. Because of the complexity of the emisssions, it is not possible to summarize the efficacy and economics of the techniques

246

adopted by the industry. Recently built petrochemical plants have incorporated in most cases modern air pollution and wastewater treatment systems. Most of the existing problems are associated with the operation of older plants, designed before strict regulations were adopted.

The installation of modern equipment for small scale plants may not be cost effective. In those cases where social and environmental benefits will result from efficient pollution control measures, economic incentives could be considered to assist in achieving compliance.

For developing countries entering the market, the application of emission performance standards is eased by the accumulated experience of earlier entrants as well as the technology innovations introduced in the field. Still, the will to apply and enforce today's strongest standards remains the most important element in environmental protection. Without enforcement, and/or economic incentives, the improvements in legislation and enactment of standards will not result in pollution abatement and prevention.

REFERENCES AND NOTES

1. Crouch, J., *Ethylene in the USA*, Occidental Chemical Corporation, Dallas, Texas, 1989.

2. Vergara, W. and Brown, D., *The New Face of The World Petrochemical Sector. Implications for Developing Countries*, World Bank Technical Paper No. 84, Washington, D.C., 1989.

3. Vergara, W., "World Petrochemicals in the Coming Decade," *Chemical Engineering Progress*, May 1989, pp. 24-32.

4. Crocco, J., "Methanol: Yesterday, Today and Tommorow," 1989 *Petrochemical Review*, Dewitt Associates, Houston, Texas, 1989.

5. Crocco, J., *Methanol Annual. An International Supply/Demand Study and Marketing Guide*, Houston, Texas, 1989.

6. Vergara, W. and Babelon, D., *The Petrochemical Industry in Developing Asia: A Review of the Current Situation and of Prospects for Development in the 1990's*, World Bank Technical Paper No. 113 (In press).

7. "Worldwide Report: Reserves Up Worldwide," *Oil and Gas Journal*, April 26, 1988, pp. 43-65.

8. Borup, M. and Middlebrooks E., *Pollution Control for the Petrochemicals Industry*, Chelsea, Mich., Lewis Publishers, 1987.

DISCLAIMER

The findings, interpretations, and conclusions expressed in this chapter are entirely those of the author and should not be attributed in any manner to the World Bank, its Board of Governors, or any of its affiliates.

8

Natural Gas as a Transportation Fuel

Robert J. Saunders
Rene Moreno, Jr.

Introduction

This chapter discusses options available for running conventional gasoline and diesel fueled vehicles on alternative fuels based on natural gas. The alternatives examined are compressed natural gas (CNG), liquefied petroleum gases, i.e. propane and butane (LPG), methanol which is an alcohol fuel derived principally from natural gas, and synthetic gasoline and diesel fuels derived from natural gas. The chapter analyzes the economics of the use of these fuels in a developing country context and compares them with gasoline and diesel fuel at various crude oil price levels. It does not discuss the use of ethanol as an automotive fuel because this is derived principally from the fermentation of sugar and is used on a large scale only in Brazil; however, many of the features of methanol use apply equally to ethanol. Neither does this chapter discuss the use of synthetic fuels derived from coal or oil shale because, although the end products are similar (gasoline, diesel fuel, methanol) the extraction, processing, and economics of the production of liquid fuels from solid raw materials differ widely from those derived from natural gas.

Compressed Natural Gas (CNG)

Natural gas is a good fuel for internal combustion engines. However, for motor vehicles it has the disadvantage of low energy density. A tank full of gasoline contains 900 times the energy content of the same volume of natural gas. For use as an automotive fuel natural gas is mixed with air and ignited in the engine cylinder by a spark or glow plug. Its physical characteristics are such that it cannot be ignited by compression alone in a conventional diesel engine. Such an engine therefore has to be converted to spark ignition or to use diesel fuel to ignite the gas. While stationary internal combustion engines can be coupled directly to a gas distribution system, the use of natural gas as a vehicle fuel requires that it be carried in a closed container on board the vehicle. In order to obtain a reasonable

249

quantity of fuel in a small volume it is necessary to compress the gas to 160-180 bar (2,300-2,600 psi) and store it in a heavy steel cylinder which serves as the fuel tank.

A Compressed Natural Gas (CNG) fueled vehicle must carry a load of roughly 2.8 kg for every liter of gasoline equivalent, while a liter of gasoline weighs only about 0.75 kg. There is thus a substantial weight penalty involved in the use of CNG which, in the case of a passenger car with a 75 liter gasoline tank, would amount to about 150 kg. This is significant in the case of small passenger cars or light delivery trucks. The weight penalty can be reduced by carrying less fuel in the CNG case although this results in a reduced range of operation. In the case of a light vehicle carrying 90 kg of CNG cylinders, the distance traveled before refueling would be of the order of 120 km. The same vehicle running on gasoline could travel some 300 kilometers before refueling. In the case of heavy trucks and buses the weight and volume penalty is not quite so serious because the CNG cylinders can be fixed along the chassis members below the load space. There is, however, still a substantial reduction in the distance which can be traveled on CNG as compared with diesel fuel. A recent development is the use of composite cylinders, sometimes of aluminum strengthened by a sheath of high tensile material. These are lighter than solid steel cylinders, but are not accepted by all regulatory authorities.

CNG vehicles are refueled by means of a high-pressure flexible hose with a probe on one end which fits into a receptacle on the vehicle. The other end of the hose is connected to a rigid high pressure fuel line, which is fed by a compressor, or by high pressure cylinders holding CNG at 200-245 bar (2,900-3,550 psi), or by a combination of both. Numerous safety devices are incorporated into the system. A system utilizing high pressure gas storage cylinders for *quick-fill* operation has a vehicle refill time of 2 to 4 minutes. An alternative system is used to recharge the gas cylinders of fleets of vehicles which return to a central depot and are not used at night, known as *trickle-fill* or "slow-fill." In this case a much smaller compressor operating at the vehicle cylinder pressure of 165 bar (2,400 psi) with little or no storage capacity can be used, to which all the vehicles are coupled simultaneously. It is therefore less costly than the quick-fill station, which is a major expense item in the CNG system, but has the disadvantage that the vehicles are immobilized for a considerable period of time while refueling.

The filling station requires a steady supply of natural gas and this is normally provided by a pipeline. The higher the pressure in the pipeline the lower filling station compression costs will be. The preferred location for a CNG filling station is therefore at or near a trunk gas pipeline which has a pressure of about 5 bar (75 psi). A lower pressure gas distribution line (1.5 bar) (22 psi) is a feasible but less desirable location, while in certain cases it is possible to establish satellite filling stations up to 300 kilometers from the pipeline by transporting CNG cylinders on semi-trailers by road. The limitations of this latter system are economic rather than technical and are discussed later. The demands of the transport sector alone, however, are not sufficient to justify extension of the gas pipeline system. For example, a fleet of 5,000 trucks using 120 liters of diesel fuel each per day would use only 450,000 cubic meters of gas per day if converted to CNG. Therefore, in developing countries, vehicle conversion to CNG is only economical in the vicinity of a gas pipeline. Furthermore, the reduced operating range of CNG fueled vehicles compared to those operating on gasoline or diesel fuel means that a high density of filling stations is required for free-range vehicles. This tends to weight the economic parameters in favor of captive vehicle fleets which operate in a restricted area, and often return to a central location.

Operating experience with cars using CNG is fairly extensive, going back to the 1920's in Italy. No major vehicle manufacturer at present produces a standardized CNG fueled vehicle, so all vehicles now using CNG are retrofits of vehicles originally designed to use gasoline or diesel fuel. Conversion kits for spark ignition gasoline engined vehicles are fairly readily available from a number of manufacturers, the only part needing to be specifically designed for each engine model being the gas mixing device attached to the carburetor. In the case of diesel engines, while some manufactured kits are available from Italy, these are designed for an Italian diesel engine and require substantial modification before they can be used successfully on other engine models.

Several countries, among them New Zealand, Canada, and reportedly the USSR, are actively promoting the conversion of vehicles from gasoline and diesel fuels to CNG. Historically, Italy and France were most active in this field, and Italy is still one of the main sources for specialized equipment for vehicle CNG conversions. Nevertheless, the total number of vehicles running on CNG is estimated to be 400,000 worldwide, a minute proportion of the total world vehicle population, estimated to be over 200 million.

Apart from road vehicles, conversions of boats and stationary engines to operate on CNG are also feasible. In most cases these are one-off conversions of diesel engines operating in a dual-fuel mode. In the case of boats, it is necessary that the range be limited and that the boat can repeatedly return to the refueling point at frequent intervals. Short link ferry service is an ideal application of CNG to boat propulsion provided that natural gas is readily available at one end of the ferry run. Inshore fishing fleets are also suitable for CNG conversion.

Environmental and safety aspects of CNG are generally good although there is a definite hazard in any operation involving high pressure gas. CNG fuel has negligible sulfur content (a small quantity of sulfur bearing compounds is added to the gas to give it a perceptible odor), and the products of methane combustion are carbon dioxide and water. There are no particulate emissions, and nitrogen oxides are only about 60% of a comparable gasoline engined vehicle. Unburned hydrocarbons emissions are higher than with a gasoline engine, but most of this is unburned methane. Non-methane hydrocarbon emissions are substantially lower than from a gasoline engine. Results are more or less similar in the case of CNG fueled diesels, with elimination of particulates (smoke) in "dedicated" CNG vehicles. Most of the smoke from a CNG fueled diesel engine occurs when diesel fuel is used for an ignition charge. In the case of spark ignited CNG diesel engine conversions, using only CNG as a fuel, nitrogen oxide emission is higher than a diesel at full load, but less at three-quarters load or lower. Particulate emissions are virtually eliminated in spark-ignition gas engines.

Liquefied Natural Gas (LNG)

The principal use of LNG is as a method of transporting natural gas from areas where it is abundant and cheap, to industrialized countries. Japan is the largest importer at the present time, followed by Western Europe and the US. LNG is also used by gas utility companies as a method of meeting peak demand by allowing storage of gas in liquid form near consuming areas. The use of LNG as a fuel is mainly confined to ships transporting it, which use boil-off gas from the LNG to fuel their engines. The use of LNG as a vehicle fuel is confined to a small number of experimental vehicles at present, and the general consensus is that the practical problems involved in using a liquid fuel at a temperature of $-160^{\circ}C$ in a road vehicle are such that they are unlikely to be in general use in the near future. LNG can readily be transported in bulk in road vehicles, and this is sometimes used as a gas distribution method for areas where demand is not sufficient to justify a pipeline connection. However, given the high cost of gas liquefaction facilities and of cryogenic containers, the economics of this for vehicle fuel supply are not acceptable at today's prices. Experiments are being carried out using LNG as a fuel for railway locomotives, and as a fuel for marine diesel engines.

Liquefied Petroleum Gases (LPG)

LPG has been successfully used as a vehicle fuel for many years and there are more than 1 million LPG fueled vehicles in operation world-wide. Because the exhaust gases of vehicles burning LPG are very low in toxic components they find a special application in enclosed spaces such as warehouses, and their environmental impact is much less than for gasoline and diesel fueled vehicles. The greatest problem with LPG is that the vapor is heavier than air and readily forms explosive mixtures with air in poorly ventilated spaces if there are any leaks in the fuel system and may form pools in low-lying areas, or in drains or sewers. Nevertheless, it has been, and is, widely used as a vehicle fuel, for example in Japan and Thailand. Provided adequate safety precautions are taken in workshops and refueling stations, LPG is a relatively safe automobile fuel.

The equipment used for an LPG vehicle conversion is basically similar to that used for CNG conversions, but since LPG is readily reduced to a liquid by compression to 8 bar (120 psi) at ambient temperatures it can be stored in a much lighter cylinder than CNG, and all the associated equipment can be of lighter gauge for the same reason. Because LPG is a liquid in the storage tank, a tank of any given size holds 5 times the energy of the same sized CNG cylinder, and the range and weight penalties with LPG are very much less than those encountered with CNG conversions as compared with a standard gasoline fueled vehicle. LPG is also readily transported in the liquid form, both by land and sea.

LPG has a very high octane rating and can be used in high compression ratio engines (i.e. diesel engines) but will not ignite by compression alone. In this respect it is similar to CNG. The possible technical options for the use of LPG as a vehicle fuel are therefore similar to those of CNG, i.e. dual-fuel mode in modified gasoline engines, dual-fuel mode in diesel engines using 15% to 20% diesel fuel as an igniter, and single-fuel mode in diesel engines modified for spark ignition. In practice, about 1.5 liters of propane (the principal constituent of LPG) replace 1 liter of gasoline and about 1.7 liters of propane replace 1 liter of diesel fuel. As a result, the fuel tank volume of an LPG fueled vehicle needs to be increased relative to a diesel or gasoline fueled vehicle to obtain the same operating range. Power loss resulting from conversion of a gasoline engine to LPG is negligible, about 5%.

Methanol

Methanol is a clear, colorless, inflammable liquid having about half the energy content of gasoline or diesel fuel for a given volume. It has a long, if sporadic, history of use as a vehicle fuel and was used as such in the earliest internal combustion engines. Subsequently, it was rapidly replaced by the much cheaper petroleum derivatives. It was used as a vehicle fuel in Germany during the Second World War, and is still used to fuel modern racing cars. It has a number of desirable characteristics as an automotive fuel, among which are its high octane rating, cool burning characteristics, and low levels of noxious exhaust emissions. In recent years there has been a revival of interest in the potential for using methanol as an automotive fuel for a number of different reasons. Among these are the fact that it can be produced from a wide range of raw materials including coal and natural gas and in large enough quantities to make it a viable alternative to petroleum based fuels in physical terms. Its low level of noxious emissions makes its use attractive in places where atmospheric pollution is a serious problem, and its potential for fueling lightweight high performance engines is of interest to automobile manufacturers.

One problem associated with the use of methanol is its toxicity, which although it is in fact no greater than that of leaded gasoline, is somewhat more hazardous since it can be mixed with water and sold as ethanol (potable alcohol). This can be overcome by the use of additives, such as 5% to 10% of gasoline, or other substances which make it unpalatable but do not affect its use as a fuel. Methanol is a colorless liquid with little perceptible odor, which is miscible with water in any proportion. This requires that greater precautions be taken in storage and handling of methanol, to avoid its contamination with water, than is the case with petroleum fuels which are immiscible with water. In actual practice water contamination up to 15% does not affect the fuel properties of pure methanol, but in blends with other fuels, more than a small percentage of water contamination causes the two fuels to separate out unless an additional co-solvent is used. Methanol has about half the energy content of an equivalent volume of gasoline or diesel fuel so that any vehicle designed to run on methanol requires twice as much fuel tank volume to obtain the same range. Pure methanol burns with an almost invisible flame, which causes problems in the case of methanol fires in sunlight. This can also be overcome by the use of additives, notably 5% to 10% gasoline, which render the flame luminous. Methanol fires, however, can be extinguished with water, which is not the case with petroleum products.

Because of its high octane number, methanol needs some form of ignition when used in a vehicle engine. In the case of converted gasoline engines the original spark ignition system is used. In the case of diesel engines, since diesel fuel and methanol are immiscible it is not possible to use the combination fuel system used with CNG and LPG, so the engine must be converted to 100% methanol and fitted with either spark ignition or a glow-plug. In the case of methanol fueled engines the dual-fuel capability, which is a feature of CNG and LPG conversions, is not possible, although it is reported that one manufacturer is testing a system in which an optical sensor can detect the difference between methanol and gasoline, and adjust the engine to run on either fuel. This would permit the vehicle to be filled indiscriminately with gasoline or methanol and to run on either. So far, this system has not got beyond the experimental stage.

The problems encountered in converting existing vehicles to run on methanol stem mainly from its chemical reactivity. The parts of the existing fuel system, in the case of a gasoline engine, which may be affected are the lining of the fuel tank (in some cases only), certain types of elastomer seals, washers, and tubing, and the metal composing the body of the carburetor. Conversion of the vehicle to run on methanol requires the replacement of these parts with methanol resistant materials and electroplating the interior of the metal body of the carburetor with nickel. In general, methanol reacts chemically with aluminum, copper, zinc, and alloys such as brass containing these materials, which must therefore be eliminated from the fuel supply system, or protected from contact with methanol by electroplating with resistant metals. If methanol becomes contaminated with water it also becomes corrosive to steel. None of these problems are as serious in practice as they sound, and technical solutions have been found to all of them.

Operating experience with modern methanol fueled vehicles is enough that most major vehicle manufacturers would consider producing them if a market were to appear, sufficient to establish a production line. In the case of cars and light trucks this would be around 100,000 vehicles per year, while for buses and heavy trucks the figure is in the range of 5,000-10,000 vehicles per year. In the United States recent revised clean-air regulations have in fact brought about suggestions that methanol fueled vehicles should be produced on a large scale.

Synthetic Gasoline and Diesel Fuels

These can be manufactured from natural gas by two production techniques; one requires the production of methanol as an intermediate step, the other uses a Fischer-Tropsch catalytic reaction to produce hydrocarbons directly from a gas composed of a mixture of carbon monoxide and hydrogen. The only chance for the process to be economical is for it to be carried out in very large capital-intensive plants. At present one plant for production of synthetic gasoline from methanol has been commissioned in New Zealand.

The products from such plants can be mixed with those derived from petroleum feedstocks and distributed in exactly the same way as

255

conventional petroleum fuels. No modifications to the vehicles themselves are needed for them to run on synthetic fuels, which can be blended in with similar fuels derived from crude oil.

Assumptions Underlying an Economic Comparison of Fuels

Above, the various technical options for using automotive fuels derived from natural gas were described. Now, the economic aspects of using several of these fuels will be considered by examining the costs of operating the same type of vehicle on alternative gas based fuels and on conventional gasoline or diesel fuel derived from crude oil. Since few vehicles are at present manufactured to use alternative fuels, the use of such fuels requires some modification to the vehicles after manufacture except in the case of synthetic diesel fuel and gasoline. The cost of these modifications is an important factor in assessing the economic viability of vehicles using alternative fuels. Apart from modifications to the vehicles themselves, modifications are required to the existing fuel distribution system; in the case of CNG the cost of fuel distribution is a major component in the overall cost of utilizing CNG as a vehicle fuel. Because vehicle operating costs and conditions vary widely from one country to another, and even within a country, this section examines the relative economics of using an alternative fuel compared to operating the original vehicles on conventional gasoline and diesel fuel at an unspecified hypothetical location, using typical developing country parameters. Thus, the study is not specific to any one country but the calculations could easily be modified to reflect specific country situations by inserting different parameters in the calculations. The analysis has been carried out in both a static scenario -- in which natural gas prices remain constant in the face of varying oil price level assumptions -- as well as a dynamic scenario in which gas prices track or move with petroleum prices. In order to focus on economic efficiency considerations, no account has been taken of taxation on vehicle fuels.

In computing the vehicle conversion costs to alternative fuels on a mileage basis, account has been taken only of incremental conversion costs, ignoring the first cost of the vehicle itself. To some extent this prejudices the case against the use of alternative fuels, because once demand for such vehicles reached levels which justified mass production at factory level, conversion costs would become minimal in comparison with those used in this study. However, this lands one in the classic chicken-and-egg dilemma: mass production of vehicles using alternative fuels will not occur until demand for new vehicles reaches an adequate level (10,000 to 100,000 vehicles per year, depending on type and other factors) while demand will not reach this level until consumers are satisfied that alternative fuels are readily available and that their use is economically advantageous to the individual consumer.

Among the basic assumptions used in the analysis are the following: (i) natural gas has a calorific value of 1000 Btu per cubic foot and is available from a pipeline; (ii) gasoline means leaded regular grade automotive fuel and diesel fuel (gas oil) is a standard automotive grade. Both are assumed to be produced in a world-scale export refinery processing Middle East light crude oil, with refining margins of the order of $5.00 per barrel. These assumptions are used in three scenarios: (a) natural gas is supplied from a pipeline at $1.00 per 1000 cubic feet (Mcf.); (b) natural gas is supplied at $2.00 per 1000 cubic feet; and (c) the price of natural gas changes in proportion to crude oil prices starting at $0.63 per Mcf. when oil prices are at $10 per barrel and rising to $4.40 per Mcf. when oil prices are at $70 per barrel. The final cost of gasoline and diesel fuel, used in establishing the baselines for comparative evaluations, includes a shipping charge, terminal and storage, charges, transport charges to supply filling stations, and filling station costs. These cost elements are calculated using representative developing country investment and operating cost values; and (iii) five classes of vehicles have been selected for study, as follows:

- Automobile, private use, gasoline fueled
- Automobile, fleet taxi use, gasoline fueled
- Light truck, commercial use, gasoline fueled
- Bus, public urban transport use, diesel fueled
- Heavy truck, urban/interurban use, diesel fueled

A number of factors have been taken into account in calculating the relative economics of different gas-derived fuels. The main ones outlined in Table 8.1 are:

- The cost of production of the resource
- The cost of converting the resource into a transport fuel
- The cost of transport and distribution of fuel to the ultimate consumer
- The efficiency of fuel utilization by the vehicles.
- The cost of converting a vehicle to use an alternative fuel.

257

Table 8.1. Vehicle and Fuel Specifications

	Vehicle Specifications				
	Passenger Cars	Light Trucks	Heavy Trucks	Taxis	Buses
Conventional Fuel Type	Gasoline	Gasoline	Diesel	Gasoline	Diesel
Miles per Gallon	30.4	23.5	5.2	30.4	7.1
Kms per Liter	12.9	10.0	2.2	12.9	3.0
Liters per Km	0.08	0.10	0.45	0.08	0.33
Btu per Km	2,624	3,394	16,482	2,624	12,071
Vehicle Life Years	14	12	10	14	12
Annual Mileage (Km)	12,000	19,000	22,000	96,000	60,000
Conversion Cost ($)					
CNG	1,000	1,800	4,500	1,800	4,500
LPG (Propane)	700	900	3,500	900	4,800
Methanol	350	500	3,200	500	4,300
Discount Rate on Conversion (%)	10	10	10	10	10
Annuitized Conversion Cost ($)					
CNG	136	264	732	244	660
Propane	95	132	570	122	704
Methanol	48	73	521	68	631

	Fuel Specifications				
	Gasoline	Diesel	CNG	Propane	Methanol
Mega Joules per Kg	47.70	44.70	-	45.70	20.02
Btu per Cubic Foot	-	-	1000	-	-
Specific Gravity (Kg/Liter)	0.75	0.86	-	0.51	0.80
Mega Joules per Liter	35.78	38.44	-	23.31	16.02

Assumed
  Crude Oil ($/Barrel) 20.00
  Natural Gas ($/Mcf) 1.00

Delivered Fuel Prices

Local	Gasoline	Diesel	CNG	Propane	Methanol
0-5 Km ($/Liter)	0.159	0.166	-	0.090	0.148
0-5 Km ($/MMBtu)	4.69	4.56	4.95	4.09	9.72
Long Distance					
300 Km ($/Liter)	0.182	0.189	-	0.115	0.171
300 Km ($/MMBtu)	5.37	5.19	7.52	5.22	11.24

Analysis of CNG

Storage and filling stations represent a major cost item in the CNG fuel system for vehicles. The major components are compressors, storage cylinders, high pressure piping, and refueling bays. Fast-fill stations are the most costly because they require multiple compressors and banks of high pressure storage cylinders capable of handling pressures up to 126 bars (1800 psi) for first stage storage (used for preliminary fill of vehicle tanks) and 243 bars (3500 psi) for second stage storage (final fill of vehicle tanks). Such a system is capable of refilling an automobile's storage tanks in about 10 minutes.

A typical CNG filling station having a throughput of 50 Mcf of gas (at normal temperature and pressure) per day would provide fast fill refueling for about 200 to 250 vehicles per day, comprising a mixture of automobiles and light trucks. 150 Mcf/day would provide for a mix of 20 urban buses, 40 taxis, and 10 light trucks or vans. The 50 Mcf/day station would cost about $150,000 and the 150 Mcf/day station about $450,000. Based on a 60% load factor nominal capacity of these stations would be 80 Mcf/day and 240 Mcf/day respectively. Table 8.2 shows delivered fuel cost under the $1.00/Mcf natural gas cost scenario.

It can be seen from these delivered fuel costs that capital and operating costs are more significant than feedstock cost. With higher gas prices than $1.00 per Mcf the differential would be less, but capital and operating costs are always a substantial part of the delivered cost of CNG where a fast-fill refueling system is employed. A fleet fast-fill refueling station would have similar capital charges but a lower operating cost, so that the delivered cost of CNG in this case would be about $3.70 per Mcf. The lowest cost supply would be a captive vehicle fleet using a trickle-fill

Table 8.2. 80 Mcf/Day Nominal Capacity CNG Fueling Station

Throughput (60% Load Factor, 310 Days/Year Operation)	15,500 Mcf
Annuitized Capital Cost @ 10% p.a. Over 5 Years	39,570
Capital Charges per Mcf	2.55
Operating Cost per Mcf	1.40
Feedstock Cost per Mcf	1.00
Delivered Cost of CNG per Mcf	4.95

Table 8.3. Delivered Cost of CNG at Remote Satellite Stations ($/Mcf)

Feedstock Cost	1.00
Compression Cost	1.53
Transport Cost	
50 Km	0.525
100 Km	0.93
300 Km	2.57
Receiving and Refueling Station	2.42
Delivered Cost to Consumer	
50 Km	5.48
100 Km	5.88
300 Km	7.52

system (no high pressure gas storage) which would have a delivered cost of about $3.30 per Mcf.

The 80 Mcf/day nominal capacity refueling station cost is the one which has been used for comparative purposes with other fuels, although it is clearly not the optimum case, which would be captive fleet refueling with trickle charging. The problem in attempting any comparison between competing fuels is that there are so many variables in each case that it is difficult to ensure that the comparisons are valid. In the case of this study, all comparisons have been made on the basis of direct sales to individual vehicle owners.

The distance that CNG is transported from its source to the filling station also affects the final cost. CNG can be transported in tube trailers a considerable distance from the main gas pipeline. Any limitations of so doing are economic rather than technical. Table 8.3 gives estimates of the cost of compressing, transporting, and refueling charges for such a system.

From the foregoing, the delivered cost of CNG to the consumer is as shown in Table 8.4.

Analysis of LPG

For the purposes of this analysis it is assumed that LPG is extracted from a rich wet gas stream. In small to medium quantities LPG is transported by road or rail in pressurized tanks. In large quantities it is moved by pipeline or by tank barge or ship. For moderate quantities shipment up to 300 km is usually by road in tankers having a capacity of

Table 8.4. Delivered Cost of CNG ($/Mcf) and Comparative Cost of Gasoline and Diesel (¢/Liter) (Calorific Basis)

Operation	CNG Cost	Gasoline Cost
Trickle-Fill Fleet	3.31	-
Fast-Fill Fleet	3.70	-
Fast-Fill Public	4.95	15.89
Base Case		
50 Km Distant	5.48	16.27
100 Km Distant	5.88	16.65
300 Km Distant	7.52	18.20

about 3400 US gallons (13,000 liters) which cost about $70,000 each. Transport costs, in cents per liter, of using such a tank-truck are as follows:

0-5 km (local delivery)	0.24
50 km	0.66
100 km	1.07
300 km	2.72

An LPG filling station handling 2000 liters/day would cost around $40,000 for tanks, flow lines, vehicle refueling equipment, and construction. Filling station costs, in cents per liter, are as follows:

capital charges	1.7
labor	1.2
miscellaneous	0.8
total	3.7

The delivered cost of LPG to the automotive consumer is shown in Table 8.5.

Analysis of Methanol

Almost all methanol is produced from natural gas. The process is somewhat similar to the manufacture of nitrogenous fertilizer, and plants are on the same scale, with similar capital costs. A typical world-scale plant would be around 1200 tons per day capacity or larger and would require an investment of $200 to $250 million. At a 10% discount rate and natural gas cost of $1.00 per Mcf, such a plant would produce methanol at

Table 8.5. Delivered Cost of LPG to Automotive Customer Based on Natural Gas Price of $1.00/Mcf

Distance from Source of Supply (Km)	Fuel Cost (¢/Liter)	Transport Cost (¢/Liter)	Station Cost (¢/Liter)	Total Cost (¢/Liter)
0-5	5.1	0.2	3.7	9.0
50	5.1	0.7	3.7	9.5
100	5.1	1.1	3.7	9.9
300	5.1	2.7	3.7	11.5

about $150 per ton. Since methanol has only half the calorific value of gasoline or diesel, this would correspond to conventional petroleum fuels at $340 per ton. The price of the product from such capital intensive plants is extremely sensitive to the rate of interest charged on capital. A paper submitted to the 3rd ASCOPE Conference in Kuala Lumpur in December 1985 by Shell Group personnel cited the case of a 2,500 ton per day plant costing $500 million and a feedstock cost of $1.00 per Mcf. At 20% capital charges the price of the product was $185/ton. There are also

Table 8.6. Methanol "Green-Field" Plant Production Cost in a Developing Country

Assumptions		
Capacity (Tons/Day)	1,200	2,500
Service Factor (%)	81	81
Production (Tons/Year)	354,780	739,125
Capital 1986 (US $ Million)	207	349
Natural Gas Feedstock and Fuel Use (Mcf/Ton)	32	32
Natural Gas Value ($/Mcf)	1.00	1.00
Operating, Maintenance ($ Million/Year)	14.0	26.0
Cost Summary per Ton Methanol		
Capital 15% DCF, 15 Year Life	76.71	62.08
Natural Gas Feed and Fuel	32.00	32.00
Operating, Maintenance	39.46	35.18
Total, Plant Gate	148.17	129.26

appreciable economies of scale. The sensitivity of product cost to feedstock prices is not as great as it is to interest charged on capital. Owing to a worldwide surplus of production capacity, international prices for bulk methanol are currently depressed. In early 1986 the price for fuel grade methanol was around $140 per ton (11¢ per liter). It is highly unlikely that a newly constructed plant in a developing country could produce methanol at such a low price (see Table 8.6). However, for the purposes of this study methanol fuel was assumed to have a landed value of $140 per ton.

Fuel methanol is generally blended with about 10% of light gasoline for a number of reasons. This product is known as M90. Since M90 is not available commercially it would be necessary to mix it at the bulk storage plant. The cost of the blend, in cents per liter, would be as follows:

methanol	10.7
gasoline	1.6
blending costs	1.1
total M90 cost	13.4

Table 8.7 shows the delivered cost of M90.

Since cars designed for methanol fuel are not generally manufactured, it is necessary to modify the vehicle fuel supply system. The cost of these modifications ranges from about $350 for a personal automobile to about $4300 for a bus. The M90 methanol/gasoline blend is used for spark-ignition gasoline engine conversions. For buses and trucks where diesel engines are converted, 100% methanol (M100) would be used as the fuel in order to take advantage of the higher thermal efficiency offered by the greater compression ratio. Although blending costs are eliminated by the use of M100, it entails greater transport and handling charges because of the need to avoid the presence of an explosive air/vapor mixture above the liquid surface in tanks, and to install anti-combustion baffles to prevent

Table 8.7. Delivered Cost of Methanol M90 Blend to Consumer (¢/Liter)

Distance from Supply Point	Blend Cost	Transport Cost	Station Cost	Delivered Cost
0-5 Km	13.4	0.164	2.576	16.14
50 Km	13.4	0.544	2.576	16.52
100 Km	13.4	0.924	2.576	16.90
300 Km	13.4	2.476	2.576	18.45

accidental ignition. Thus the retail cost to the consumer of M90 and M100 would be approximately the same.

Synthetic Gasoline and Diesel Fuel from Natural Gas

The technical feasibility of manufacturing synthetic gasoline and diesel fuel from natural gas is well established, however, economic factors rule out any application of these technologies at present-day prices for crude oil. Table 8.8 gives a summary of estimated capital and operating costs for plants to produce these fuels from natural gas. As in the case of methanol, these plants are highly capital intensive and the price of the end product is very sensitive to interest rates charged on capital. The capital costs shown in Table 8.8 are much higher than these given in much of the published literature, on the basis of the very high level of cost overrun experienced, for example, in construction of the New Zealand synthetic gasoline plant, which uses the Mobil process with methanol as an intermediate product. This is at present the only large scale plant manufacturing gasoline from natural gas. Because of the very high cost of

Table 8.8. Synthetic Gasoline and Diesel "Green-Field" Plant Production Cost in a Developing Country

	Gasoline (Mobil Process)	Diesel (Fischer Tropsch)
Assumptions		
Capacity (Tons/Day)	1,775	1,775
Service Factor (%)	81	81
Production (Tons/Year)	524,779	524,779
Capital (1986 US$ Million)	1,460	1,550
Natural Gas Feedstock and		
Fuel Use (Mcf/Ton)	85.2	71.6
Natural Gas Value ($/Mcf)	1.00	1.00
Operating, Maintenance		
($ Million/Year)	90.0	95.0
Cost per Ton Synthetic Fuel ($/Ton)		
Capital 10% DCF, 15 Year Life	366	388
Natural Gas Feed and Fuel	85	72
Operating, Maintenance	172	181
Total, Plant Gate	622	641

synthetic gasoline, the New Zealand government is still encouraging conversion of automobiles to CNG and LPG fuels in order to limit overall consumption of gasoline in the country. Despite the high cost of these synthetic fuels, the existence of these technologies imposes an upper limit to the price which crude oil can command as a source of transport fuels. Furthermore, advances in catalytic conversion technology may, in the next decade or so, substantially reduce the cost of the necessary plant below what is required using present day technology.

An Economic Comparison of Fuels

Based on the assumptions outlined earlier, Tables 8.9 through 8.11 summarize the break-even costs of crude oil at which the overall vehicle cost of using selected natural gas based fuels would be the same as for conventional fuels. This analysis is shown both at the point of bulk supply (0 km) and at refueling points 300 kms distant from the point of bulk supply. In Tables 8.9 and 8.10, the prices of alternative fuels are assumed not to be affected by changes in crude oil price (i.e. the price of natural gas is assumed fixed at $1/Mcf and $2/Mcf respectively), while in Table 8.11, natural gas prices are assumed to "track" crude oil prices proportionately.

Table 8.9. Break-Even Crude Oil Price Levels (US$/Bbl), Natural Gas Price Constant at $1.00/Mcf

	CNG Fast-Fill	CNG Trickle-Fill	LPG	Methanol (M90)
Local Delivery (0-5 Km)				
Cars	45	36	33	65
Taxis	27	18	19	57
Light Trucks	44	35	28	63
Heavy Trucks	33	24	26	68
Buses	27	18	23	63
Distant Delivery (300 Km)				
Cars	55	46	35	71
Taxis	37	28	22	63
Light Trucks	54	45	30	69
Heavy Trucks	44	35	29	74
Buses	38	29	26	70

Table 8.10. Break-Even Crude Oil Price Levels (US$/Bbl), Natural Gas Price Constant at $2.00/Mcf

	CNG Fast-Fill	CNG Trickle-Fill	LPG	Methanol (M90)
Local Delivery (0-5 Km)				
Cars	50	41	46	76
Taxis	32	23	32	67
Light Trucks	49	40	40	73
Heavy Trucks	39	30	39	78
Buses	33	24	36	74
Distant Delivery (300 Km)				
Cars	60	51	48	82
Taxis	42	33	34	73
Light Trucks	59	50	43	79
Heavy Trucks	50	41	42	85
Buses	44	34	38	81

Table 8.11. Break-Even Crude Oil Price Levels (US$/Bbl), Natural Gas Price "Tracks" Crude Oil Price

	CNG Fast-Fill	CNG Trickle-Fill	LPG	Methanol (M90)
Local Delivery (0-5 Km)				
Cars	59	46	93	161
Taxis	32	19	31	136
Light Trucks	57	44	69	153
Heavy Trucks	43	29	67	178
Buses	33	19	50	165
Distant Delivery (300 Km)				
Cars	75	61	104	178
Taxis	47	34	42	153
Light Trucks	73	60	80	170
Heavy Trucks	59	45	80	198
Buses	50	36	64	186

Conclusions

The analysis indicates that *CNG Fast-Fill* requires oil prices to rise to $27 per barrel before it becomes competitive with gasoline for taxis and bus fleets, even if gas prices were assumed to remain constant at the low level of $ 1.00 per Mcf. In contrast, for these vehicles, *CNG Trickle-Fill* would become competitive at $18 per barrel. Even when natural gas prices are allowed to track oil prices, CNG Trickle-Fill would be competitive at oil prices of $19 per barrel or higher.

When natural gas prices are kept constant at $1.00 per Mcf., *LPG* becomes competitive at $19 per barrel, when used in taxis. However, if gas prices are kept fixed at $2.00 per Mcf. or are allowed to track oil prices, crude oil prices would have to reach a considerably higher level, about $32 per barrel, before LPG competes economically.

Finally, *methanol* is seen to be the least competitive natural gas based option. Even in the most favorable circumstances, oil prices would have to reach $57 per barrel before methanol is considered as attractive as gasoline.

A sensitivity analysis with respect to the basic parameters shows that for annual distance traveled, costs increase rapidly as the distance traveled falls below the base case used in the study, but decreases much less for distances above the base case. For other parameters such as conversion cost and interest rates, the rate of change is linear and directly proportional to the amount of change above or below the base case. This analysis explains the difference, for instance, in the operating economics of taxis relative to private cars. Although these are basically the same vehicle, the much higher annual mileage traveled by the taxi makes the use of alternative fuels economically attractive at crude prices well below those needed to justify their use in private cars.

The conclusions to be drawn from this analysis are that at crude oil prices of $10 per barrel and lower, alternative fuels are generally uncompetitive. Between $10 and $20 per barrel custom built propane fueled high mileage vehicles and retrofitted vehicles using CNG trickle-fill refueling, become competitive. Between $20 and $30 per barrel, CNG fast fill and propane fueled low mileage vehicles could be competitive. Methanol becomes competitive above $50 per barrel, while synthetic gasoline and diesel fuel do not become competitive until the crude oil price reaches levels above $70 per barrel. For CNG fueled vehicles, the high cost of transporting this fuel in tube trailers means that CNG is competitive at the crude oil price levels quoted above only when the filling station is located at or close to a pipeline or a gas-field.

The most appropriate use for CNG is for captive vehicle fleets with a relatively high annual mileage but a restricted range, and whose duty cycle is such that they can use the trickle-fill recharge system. From this conclusion it might be deduced that large truck and bus fleets, as well as taxis operating within limited range, would fall in this category. However, trickle-fill supply systems generally require the vehicle to be inoperative over night (or, alternatively, in day time for a similar length of time). This, along with the fact that these vehicles tend to have very high utilization rates, especially in developing countries, implies that a higher level of capitalization (in terms of vehicle numbers) would be required to maintain the same level of service. Alternatively, the vehicles would have to be converted into dual use (CNG as well as gasoline or diesel) systems. In both circumstances, the effective cost would be considerably more than might appear in the initial analysis. From this perspective, trickle-fill may in fact not be appropriate for these vehicles. In addition, some research has suggested that average cost of operation of bus fleets in developing countries may in fact increase with size. Therefore, while larger size fleets may be more suited for conversion to CNG in terms of the trickle-fill system characteristics, there are other factors intrinsic to large fleets which may work in the opposite direction.

The worst case for conversion to any alternative fuel is the low mileage free-range private automobile, which explains why the New Zealand and the Canadian Governments (the two most actively involved in promoting CNG in recent years) have found it necessary to offer substantial subsidies to private motorists, in order to persuade them to convert to CNG.

At crude oil prices prevailing in 1987/88, there was little economic justification in converting to alternative fuels. At the same time, the evolution of the crude oil price cycle suggests that price levels in the 1990's could climb back towards the range of $20 to $30 per barrel, at which point some degree of conversion will become economically attractive. However, the unpredictability of turning points in the world oil market and the possibility of further downturns beyond the mid-1990's as well as the likely occurrence of future fluctuations, merit a prudent approach. In particular, conventional oil based transport fuel supply facilities can be expanded rather quickly and conveniently, with relatively little investment. On the other hand, a large scale conversion to gas based fuels will require significant investments on both the supply and demand sides, as well as the problems of coping with a comparatively unfamiliar technology on a grand scale. Therefore, the economic justification of large investments in gas infrastructure in an uncertain climate, will probably require net conversion benefits that are above the marginal gains which can be derived at $20 to $30 per barrel of crude oil. It will also require that oil prices remain at these levels long enough to sufficiently reinforce expectations of continued high future prices.

It is clear that at crude oil price levels of $60 per barrel and above, massive substitution of conventional gasoline and diesel fuel by more abundant alternative fuels is likely to occur. Since the transport sector is the only one in which no viable alternative to oil based fuels existed in the 1970s, the existence of substitute fuels implies that crude oil price levels above $60 per barrel (in 1986 dollars) are not sustainable in the long term, and improvements in alternative fuel technology may even reduce this upper bound.

While from a purely economic perspective, the present analysis suggests that major new initiatives to use alternative transport fuels derived from natural gas cannot be justified given present technology and assumptions about oil price levels, the benefits of natural gas based fuels goes beyond their potential as an economic substitute for oil based fuels. It is well known that transport vehicle emissions resulting from the burning of oil based fuels contain significant localized pollutants as well as elements which may contribute to the now well publicized "greenhouse effect." Serious consequences have been forecast by some meteorologists and ecological experts, if the release of these emissions into the atmosphere is not contained. In this light, the value of natural gas based fuels could become significantly more than simply the amount of conventional fuel displaced by their use. Consequently, if significant pollution related externalities are identified, the crude oil price levels at which natural gas based fuels would begin to be competitive, could be lower than those cited here. At the same time, the impact on the atmosphere stemming from vehicle emissions, as compared to that from industrial and power generating plants, would be relatively small. Hence, the potential overall environmental impact resulting from using natural gas based fuels in vehicles would probably be limited.

Finally, it must be emphasized that the above conclusions are based on existing technology. The development of low pressure absorption systems for CNG, as used for example with industrial acetylene gas, could lower the cost of using CNG. Improved chemical processes, particularly the development of new catalysts, could lower the cost of producing methanol, synthetic gasoline, and diesel fuel.

BIBLIOGRAPHY

Moreno, Rene Jr. and D.G. Fallen Bailey. *Alternative Transport Fuels from Natural Gas.* World Bank Technical Paper Number 98, 1989.

Alternative Fuels for Use in Internal Combustion Engines. Energy Department Paper Number 4, The World Bank, November 1981.

Julius, DeAnne and Afsaneh Mashayekhi. *The Economics of Natural Gas: Pricing, Planning and Policy.* Oxford University Press, 1990.

World Petroleum Markets. World Bank Technical paper Number 92, 1988.

Churchill, Anthony A. and Robert J. Saunders. *Financing of the Energy Sector in Developing Countries.* Industry and Energy Department Energy Series Working Paper Number 14, The World Bank, April 1989.

Mashayekhi, Afsaneh. *LNG Export Opportunities for Developing Countries and the Economic Value of Natural Gas in LNG Export.* Energy Department Paper Number 11, The World Bank, October 1983.

ESMAP, *Greenhouse Gases and the Potential for Global Warming.* Industry and Energy Department, The World Bank, November 1989

McKeough, Kay. *Current International Gas Trades and Prices.* Industry and Energy Department Energy Series Working Paper Number 9, The World Bank, November 1988.

Chesher, Andrew and Robert Harrison. *Vehicle Operating Costs.* The Johns Hopkins University Press, Baltimore, 1987.

ACKNOWLEDGMENTS

This chapter is based on work undertaken by the authors as well as significant contributions by current and former colleagues including D.G. Fallen Bailey, Anthony A. Churchill, Afsaneh Masheyekhi, Pierre Moulin, and John Lowe. The World Bank/UNDP/Bilateral Aid Energy Sector Management Assistance Program supported the final analysis contained in this paper. The findings, interpretations and conclusions expressed in this paper are entirely those of the authors and should not be attributed in any manner to the World Bank or its affiliates, or to the ESMAP program. Major portions of this work have been previously published as World Bank Technical Paper Number 98.

9

Natural Gas - Interchangeability with Other Fuels

Carl W. Hall

Introduction

While the use of natural gas has been increasing for many decades, the demise in its use as a major fuel has been predicted. The prediction is based on the premise that being a limited resource natural gas was being consumed more rapidly than discovered. It is reasonably well accepted that a limited resource was being consumed more rapidly than natural gas was being formed. There is a limited amount of natural gas available, but it is impossible to predict the quantity of resources in reserve. A surprising development is the recent finding of tremendous quantities of natural gas in the earth, some in underdeveloped areas of the world, such as Bolivia, Argentina, Malaysia, Indonesia, China, and the USSR, and much in deeper sites in the earth's crust throughout the world. Regardless, the consumption of natural gas, as a result of its convenience and environmental safety, is increasing, suggesting that diminishing resources will be available with an attendant increase in cost. An economy built entirely on natural gas could be precarious, except that natural gas could be mixed with or succeeded by other gaseous fuels, utilizing the available distribution network to meet needs.

Uncertainties in availability of natural gas abound. These uncertainties include: uncertainty of reserves; uncertainty of availability (even though the gas is present); uncertainty of government regulations; and uncertainty of the global economy. This chapter covers the interchangeability of natural gas with other gaseous fuels to meet a wide variety of needs.

Natural Gas Characteristics

Natural gas is largely methane, CH_4, a hydrocarbon found in its natural form (80 to 97%) which occurs in porous layers of the earth, often associated with or intermixed with oil or above the oil in a reservoir. Natural gas is often found in separate gas reservoirs. Other hydrocarbons of the paraffin series, such as ethane, propane, and butane may be present in much lesser amounts. Other gases which may be present are carbon

273

dioxide, nitrogen, helium, and hydrogen sulfide. The cost of conversion of the raw natural resource to natural gas for fuel or for a chemical use is usually minimal compared to the cost of conversion to gas of other hydrocarbon fuels, such as coal or oil derivatives. (See Chapters 5 and 7.)

Natural gas, methane, has no odor, so unless sufficient hydrogen sulfide is present, or mercaptans are added, leaks could occur in the system and go undetected. Careful maintenance, control, and operation of gas equipment are required for safety in its use. Natural gas is explosive in air in concentrations between 5.3 and 14 percent (for other gases see Table 9.1); and is lethal to humans.

Natural gas, being a gaseous material, has a low density, about 600 volumes of gas is equivalent to 1 volume of liquid (Table 9.2). Transportation of natural gas across land or through short distances in water is by pipeline. In the US there is a vast network of pipelines of approximately 1 million miles connecting sources to use, and the network is expanding. Similar networks exist in many areas of the world. For instance, a new pipeline now connects Patagonia near the Antarctic Circle with the Buenos Aires area in Argentina, and a large pipeline now connects the offshore India gas fields with the inland markets. Other extensive networks exist and are being planned in Western Europe, Mideast, Northern Africa, and Indonesia. Vast storages, usually in earth cavities, such as old mines, caves, or subterranean openings, are used to store gas to meet high demands during peak periods, and to provide a uniform supply to meet needs.

Table 9.1. Flammability of Gases in Air
(Lower to Upper Range, in Percent)

Gas	Percent
Blast Furnace Gas	35-73.5
Butane	1.9-8.5
Carbon Monoxide	12.5-74.2
Coal Gas	4.8-33.5
Hydrogen	4-74.2
Hydrogen Sulfide	4.3-4.6
Methane	5.3-14
Producer Gas	20.2-71.8
Propane	2.2-9.5
Water Gas	6.9-70.5

Source: Hall & Hinman, 1983.

Table 9.2. Densities of Gases at 1 Atm and 0°C.

Gas	Lb/Ft^3	g/L
Air	0.0807	1.2929
Carbon Dioxide	0.1234	1.9769
Carbon Monoxide	0.07806	1.2504
Hydrogen	0.00561	0.0898
Methane	0.0448	0.718
Nitrogen	0.07807	1.2505
Propane	0.1254	2.009
Oxygen	0.08921	1.429

Source: Hall, 1981.

For the international shipment of natural gas, particularly over the oceans, gas is compressed and liquefied and transported on ships and barges. Gas cannot be converted to a liquid by compression only at normal atmospheric temperatures. The gas must be compressed and cooled, normally at 600 psi. Storage tanks for liquid natural gas are double-walled with insulation between.

Natural gas from different sources varies somewhat in its thermal content, a major factor to be considered when used for fuel or heating operations. The usual heating value by combustion is about 1000 Btu per cu ft (37.2 MJ per cu m). The impurities in natural gas may contribute to the heat content, but the impurities have less heat than methane. The heating values of various gaseous fuels are presented in Table 9.3.

Some impurities cause undesirable environmental impacts. To avoid excessive pollution either the impurity is removed before combustion if used for heating, or the stack gases, the products of combustion, are treated. Generally, natural gas produces much less atmospheric pollution than other hydrocarbon fuels. There is, nevertheless, a worrisome potential impact of eventual methane emissions (through leaks or incomplete combustion) on the green house effect. For example, methane traps heat on the earth surface at a rate 16 times higher than CO_2. In addition, there is no residue ash from burning natural gas left in the combustion chamber. With properly controlled combustion of natural gas, the effect on the environment is very slight, making natural gas the fuel of choice for heating. Natural gas used for chemical production is usually treated so as to provide nearly 100 percent methane.

Table 9.3. Heating Values of Dry Gases, STP, by Volume (Approximate)

Gas	Btu/Cu Ft[*]
Biogas	540-700
Butane	2900-3200
Carbon Monoxide	320
Coal Gas	120-600
Corn Cob (850°F)	400-425
Fuel Gas (pyrolysis)	100
Hydrogen	4900
Methane	1040
Municipal Wastes	350-500
Natural Gas	1000-1025
Producer Gas	130-155
Propane	2550
Synthesis Gas	300-500

Source: Hall, 1981.
[*]1 Btu/cu ft = 37.222 kJ/cu m

It is generally assumed that methane in natural gas is a product of anaerobic activity of biological materials, known as biotic activity. The methane was produced over a long period of time from deteriorated biological matter confined in enclosed spaces in the earth without access to oxygen after the original oxygen was depleted. Over time, with the pressure of the earth, the action of the microorganisms, and the heat of respiration, the organic matter was converted to natural gas. The possibility of abiotic produced methane, in which methane was trapped in the folds of the earth as it was formed, rather than being formed from biological action in the earth, is believed by some to be the source of much methane, perhaps yet to be discovered. The conditions surrounding the earth as it was being formed, similar to methane surrounding some other planets today, could have been such that considerable methane was present. As the earth was formed, this gas could have been entrapped and incorporated in the crust or envelop of the earth. Additional credence is being given to this theory as additional natural gas resources are being located at deeper locations in the earth and is being found at folds or overlay areas of the earth.

End Uses of Natural Gas

Chapters 5, 6, 7 and 8 also cover several aspects of end uses of natural gas. The principal end use worldwide for natural gas is for heating and cooling, with heating leading in importance (Figure 9.1). Other major uses of natural gas are for producing electricity, chemical feedstock, and for transportation. Heating operations can be of two kinds, direct and indirect. With direct heating the products of combustion accompany the heated air to the product to be heated. That process is the most efficient if the products of combustion can be tolerated and there is appropriate conveyance of the products of combustion to disposal.

With indirect heating a heat exchanger separately directs the heated fluid and the products of combustion, resulting in a slightly lower efficiency and higher cost of operation. These heating operations are used in home, farm, business and factory for space and product heating. Gaining in importance once again is the use of natural gas as an energy source for driving an absorption refrigeration system for cooling.

The production of electricity with natural gas is a major use. Natural gas, as a result of its cleanliness upon combustion, its convenience, and flexibility in use makes an ideal fuel for producing electricity. There is some pressure to avoid using natural gas for combustion processes and to use other fuels for these processes. However, both the economics and minimal impact on the environment attract natural gas to these stationary uses. Electricity itself can be used for power and heating, as well as for producing light and for providing control. There are several routes to producing electricity with steam. The conventional method is to use gas or fossil fuels to heat water to produce steam which is used to drive a steam turbine and generator to produce electricity. The gas can be combusted to drive a gas turbine directly which drives a generator to produce electricity. A combined cycle using gas and steam gives higher thermal efficiency than using gas alone, thus gaining in use. Gas alone is used to supply a gas turbine, in contrast to a steam turbine. The gas turbine can be brought on line quickly to meet high demands. Although electricity is an energy source that is convenient, flexible, and clean it is generally accepted that major heating operations are less expensive by using the fuel directly rather than putting it through the electric generation steps.

Natural gas is used as a fuel for some vehicles. Propane in liquid form at normal atmospheric temperatures has gained wider use than natural gas, although still of limited scope. Liquid natural gas needs to be kept at low temperatures. Transportation applications for natural gas are generally limited to vehicles used within limited or prescribed areas or routes, primarily where the gas is available within the vicinity of use. In the city of

Figure 9.1. End Uses of Natural Gas

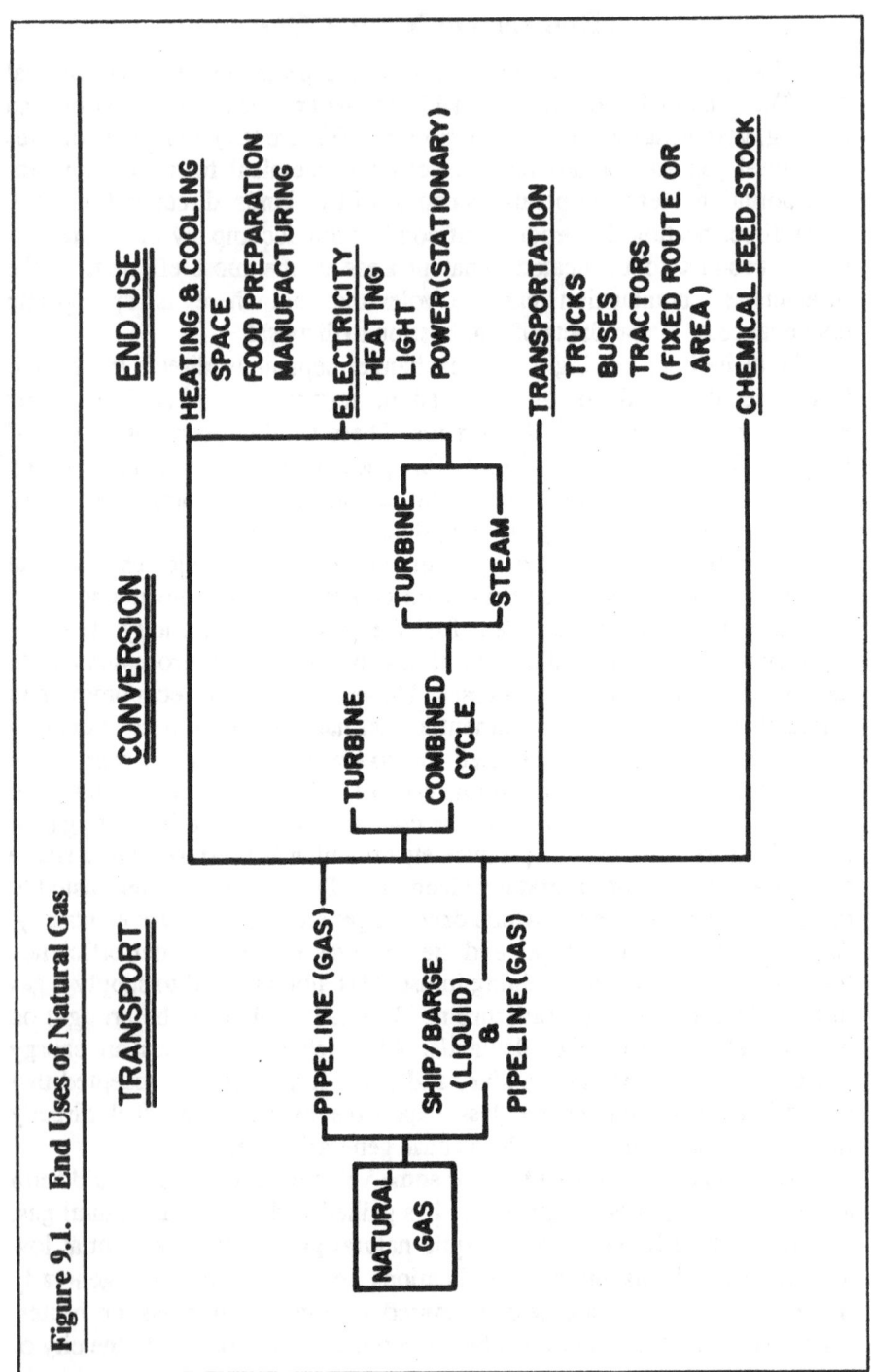

Ghengtu, Szechwan province, China, public buses with large flexible and inflatable bladders covering the roof area, hold a supply of natural gas for a prescribed route, using gas found near the city. In addition to being a convenient and available fuel, pollution problems are minimized. Similar use of gas can be made for trucks, tractors, and automobiles, but the large volume needed mitigate against extensive worldwide use at this time. If the gas is available, it seems reasonable and prudent to use it for transportation rather than to use limited foreign currency to purchase gasoline or diesel fuels.

As a chemical feedstock, natural gas is a primary source of methane. Helium may be contained in the fresh gas from the well which can be used for commercial purposes. Methanol can be produced from methane, a liquid alcohol product that can be used for many of the same purposes as natural gas. Other commercial applications for use of natural gas include production of hydrogen, which may also serve as a fuel and manufacture of ammonia. Fertilizer, plastics, and adhesives are produced from natural gas that has been reacted with steam.

Potential Natural Gas Substitutes

Hydrocarbons

Natural gas is known as a high-Btu fuel, (high-Joule) 1000 Btu/cu ft (37.3 MJ/cu m). In contrast there are low-Btu fuel (low-Joule) 100-250 Btu/cu ft (3.5-9 MJ/cu m), and medium-Btu fuel, 300-700 Btu/cu ft (13-27 MJ/cu m) (Figure 9.2). For direct substitution of natural gas it would be desirable to have its replacement contain 1000 Btu/cu ft. Most other gaseous fuels have a lower heat content, some of which can be upgraded to provide a higher heat content, but all potential substitutes are less energy intensive than natural gas at the same pressure.

The major fossil fuel resource of the world is coal. At least a 500 yr supply, based on known reserves and present rate of use, is available. Coal can be gasified to produce a gas that can be substituted for natural gas (Figure 9.3). Generally, the gas produced from coal is in the medium-Btu/cu ft range, which is called synthesis gas or coal gas. The gas is made by destructive distillation of coal, in absence of air, and consists mainly of carbon monoxide and hydrogen. Synthesis gas can be upgraded to high-Btu gas by a process called methanation, in which carbon and hydrogen are passed over appropriate nickel catalysts to produce methane. This enriched gas is called substitute natural gas. The term synthetic natural gas (SNG) is used to represent that gas produced directly from the carbon in the coal by methanation. Low-Btu gas, also called producer gas, blast furnace gas, power gas, and water gas, is primarily for industrial use

Figure 9.2. Low-, Medium-, and High-Joule Gas

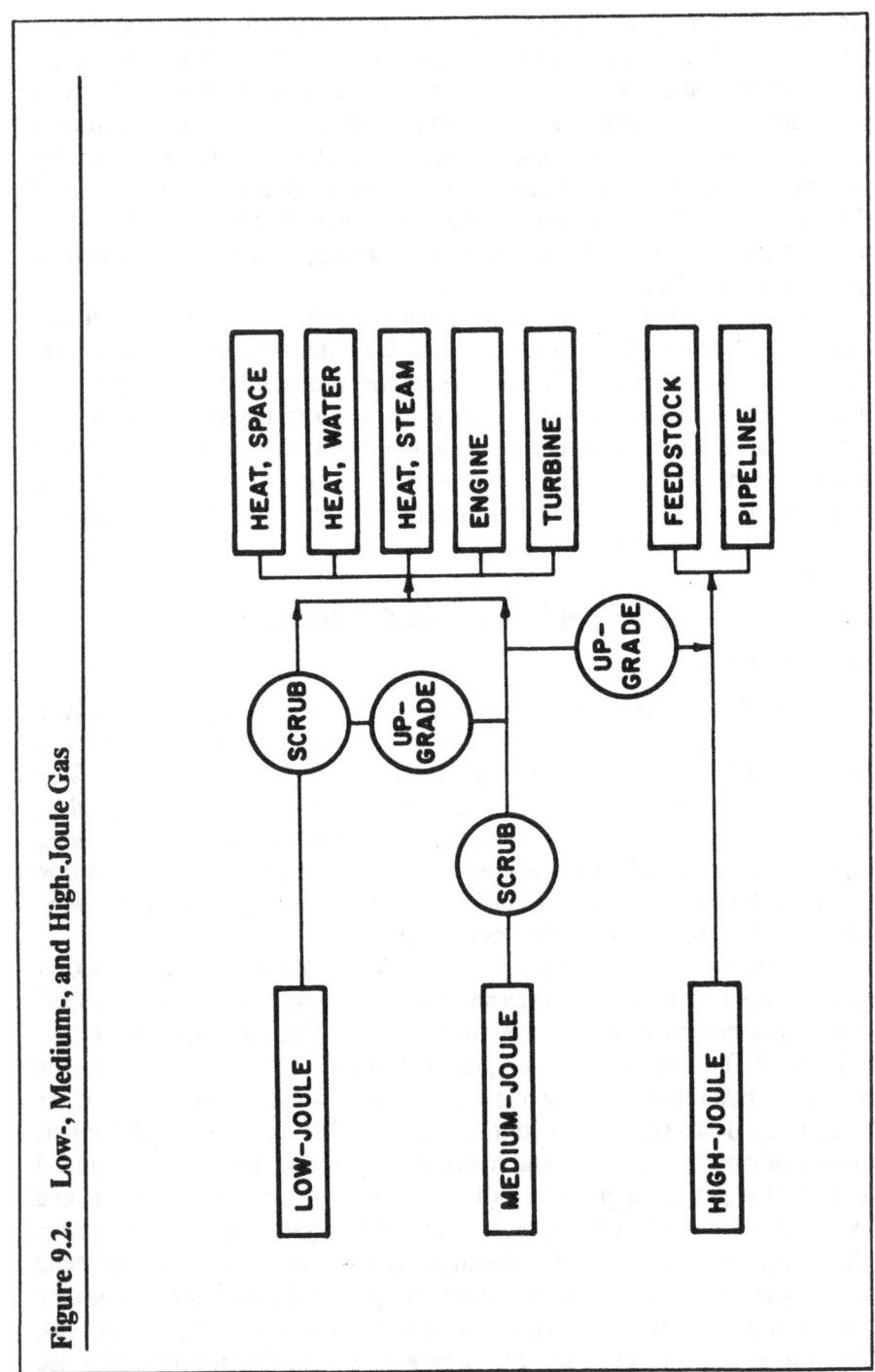

Figure 9.3. Coal Gasification

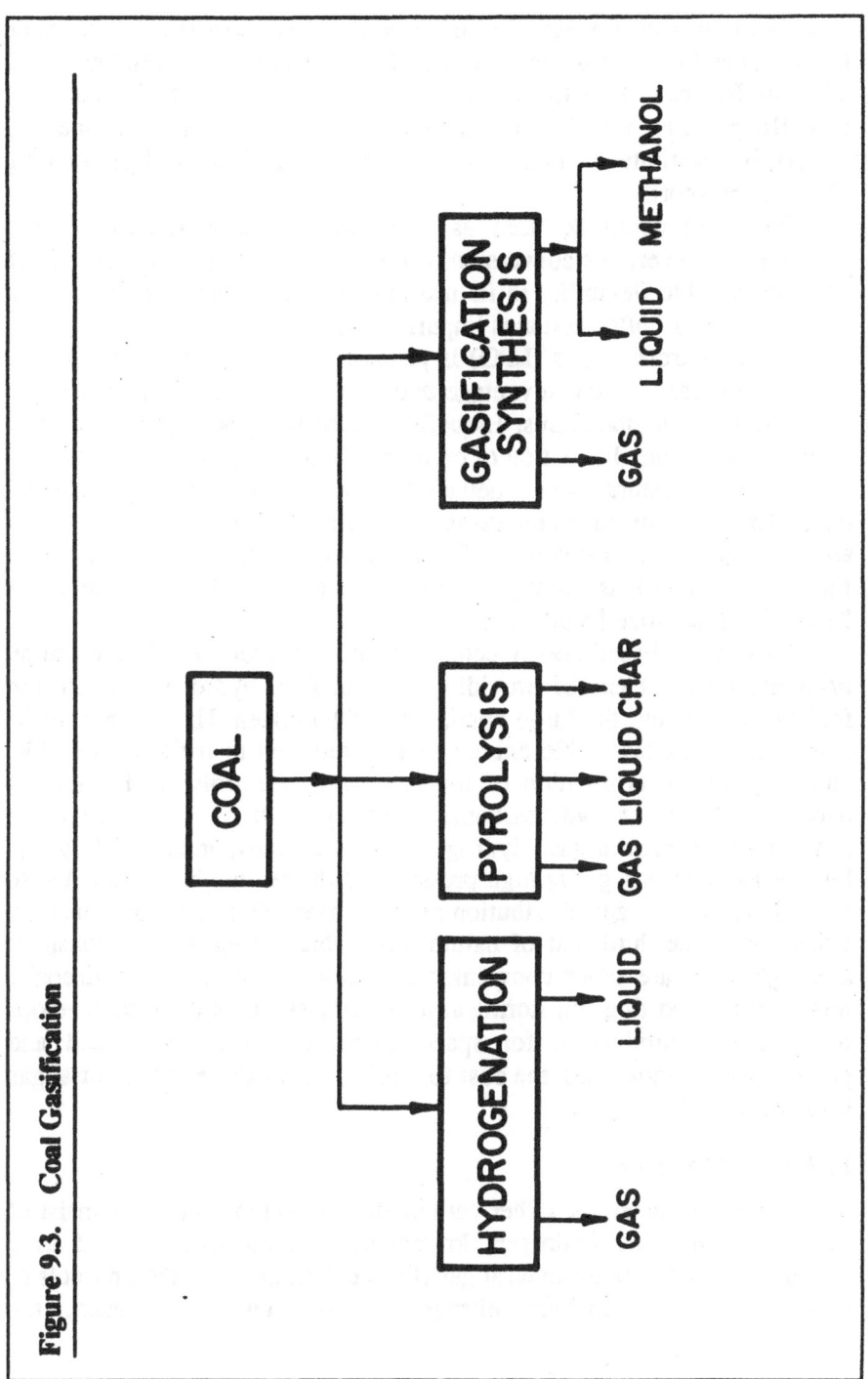

and could be used for electric power production. Low-Btu gas is made from coal and other fossil resources such as tar sands, oil sands, and shale. The low-Btu gas made from coal is about 25 percent carbon monoxide. Low-Btu gases are usually produced near the point of use and do not easily or readily substitute for natural gas and do not lend themselves for long distance transport.

Petroleum could be used as a resource from which to produce methane. However, this conversion is not usually made because petroleum has considerable flexibility in its use, being a liquid product. Petroleum, like coal and gas, often contains impurities such as sulfur.

Liquid petroleum gas (LPG) is primarily propane which, in contrast with natural gas, has the advantage that it can be liquefied by pressure at above freezing temperatures. Liquefied petroleum gas is produced from natural gas or from distillation of crude oil. Generally, the end use of LPG is similar to natural gas (methane) with the use from pressurized containers, such as for vehicles-automobiles, tractors, trucks, and buses and for heating. LPG compares favorably with other liquid and gaseous fuels, except that it is usually more expensive than natural gas, and it is handled in pressurized containers.

A hydrogen based energy economy has been proposed for the future, primarily on the basis of providing a clean fuel, hydrogen combustion forming water, and the large abundance of hydrogen. Hydrogen may be abundant, but considerable energy will be required to make it available, either by steam reformation of methane or by electrolysis of water by partial oxidation of hydrocarbons. Presently most of the hydrogen is produced from natural gas. Hydrogen is usually transported as a liquid at low pressure or as a gas at high pressure. Hydrogen might be mixed with natural gas and the gas distribution network used. Hydrogen has a heating value about one-third that of natural gas, which designs must consider, although there are other compensating factors, such as lower viscosity related to transporting and storing as a gas. Unless a high demand develops for hydrogen production for space travel, chemical uses, food and pharmaceutical industries, the cost for energy competitive with natural gas is unlikely.

Biological Materials

Biological materials, either carbohydrate or lignocellulose, consist of carbon, oxygen and hydrogen, known as biomass, can be used as a resource to substitute for natural gas (Figure 9.4). Some of the products of biomass products include nitrogen in the air. In essence, the

Figure 9.4   Biomass for Fuel

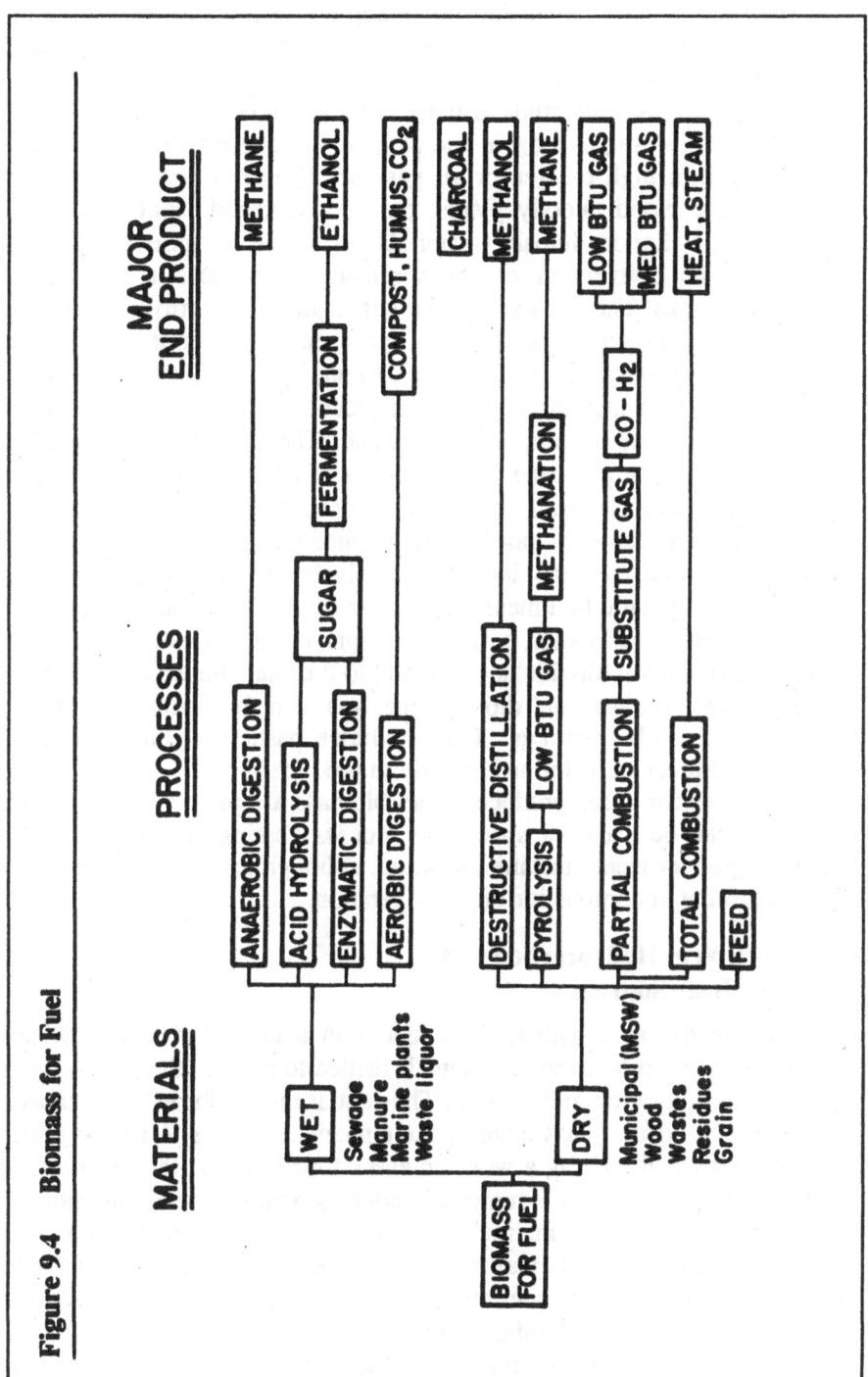

naturally-occurring conversion of biological materials in the earth's surface or crust is replaced by manmade processes including high pressure and temperature but done in a very short time.

Lignocellulosic materials, usually at a low moisture content (5-25%), consisting of carbon, hydrogen, and oxygen, are converted by gasification or heating in a high oxygen or air gasifier. The principal lignocellulose materials are wood, woody plants, paper, straw, and plant refuse. The pyrolysis of lignocellulosic materials provides a gaseous product. Generally the conversion of these materials produces a low-Btu or medium-Btu gas. The products of the gasification or pyrolysis operations are primarily carbon monoxide and hydrogen as the energy producers. Carbohydrate materials, consisting of carbon, hydrogen, and oxygen, can be converted by anaerobic digestion, a process that parallels in principle the natural conversion process to methane. The principal carbohydrates used are grains, sugar cane, and potatoes. Depending on the moisture content and characteristics of the carbohydrate materials, they can be converted to medium-Btu gas directly by anaerobic digestion. Other paths to production of methane include hydrolysis of the material by acid, thermal or enzymatic treatment prior to anaerobic digestion. The products of the anaerobic digestion process are primarily methane and carbon dioxide. The carbon dioxide does not add to the heat content of the fuel, so by removing the carbon dioxide the gas can be increased from a medium-Btu to a high-Btu gas. Carbohydrates, particularly those materials of low moisture, can be burned directly to produce heat.

End uses for gases produced from biomass are primarily for heating. The heat can be used for space and process heating, and steam boiler, steam engine, or steam turbine operation, although less often use could be for gas turbine or combustion engine operation.

Comparison of Hydrocarbon and Carbohydrate Sources

Natural gas as a hydrocarbon fuel is in a near ready state for use. Resources need to be located and wells drilled to remove the gas. A major cost is for exploration and drilling. The cost at the wellhead has increased as the depth of drilling has increased to tap new resources. A rule of thumb is that the cost of drilling a well increases four times as the depth of the well is doubled. The gas is removed under its own pressure, but pumping stations are needed to maintain uniform pressures throughout a long pipeline system. The gas may need to be treated depending on the use. Potential treatments include removal of sulfur, reduction of moisture content, and removal of other gases such as helium, depending on the source of gas. The distribution system developed for natural gas can be used for many other gases. Modifications would need to be made in metering devices, pressure, and openings (spuds) at the burners to accommodate other gases. Other gases derived from liquid or solid hydrocarbon fossil fuels vary in content and mixture of gases, heating value, density, pressure, viscosity, and safety.

Biomass as a carbohydrate fuel is dependent on solar energy as a driving force for production. In contrast to hydrocarbon fuels, carbohydrates are widley dispersed as plants and as derivatives and wastes of plants and animals. Although some varieties of plants may be grown specifically for biomass for fuel, the first phase in the utilization of biomass for fuel will be to utilize the products, byproducts, waste products, and selected elements of plants for fuel. As widely dispersed materials during growth in the field, forest, and in water, the cost of harvesting and collecting these materials can exceed the value of the product itself. Thus, wastes from food processing plants, bagasse from sugar refineries, spent hops form beer manufacture, straw and fodder from threshing, spent whey from cheese making, sawdust from lumbermills, and municipal solid wastes (MSW) (Figure 9.5), all of which concentrate the product at a site, will be used for making fuels before major crops will be harvested specifically for fuel. In these examples the product is harvested and accumulated for the primary product, and the use for fuel is a byproduct, with the costs attributed to the primary product. Considerable variability exists in biological sources of fuel-moisture content, chemical content, density, maturity, rate of growth-increasing the difficulty of maintaining quality of gas produced. The cost of fuels produced from plant materials is at least twice the cost of the readily available natural gas. With time the more expensive gas produced from biological materials and the less expensive gas available from the surface of the earth are expected to approach each other in value. When that occurs, the elaborate distribution system for natural gas can be used, with modification, for gas produced from biological materials.

For biological resources to be given serious consideration on a commercial basis as a source of gaseous fuel, the necessary plants need to be grown in the general vicinity, 100 miles or so of the location of use. Thus, temperate or tropical areas with near optimum temperature and soil and moisture provide the potential of providing a material that could be used for producing a gas that could substitute for natural gas.

Manufactured Gas

It is instructive to realize that manufactured gas preceded natural gas as a fuel and for light. In the US the manufacture of gas began in the early 1800's. Two processes predominated. The first, known as manufactured

Figure 9.5   Use of Municipal Solid Waste (MSW)

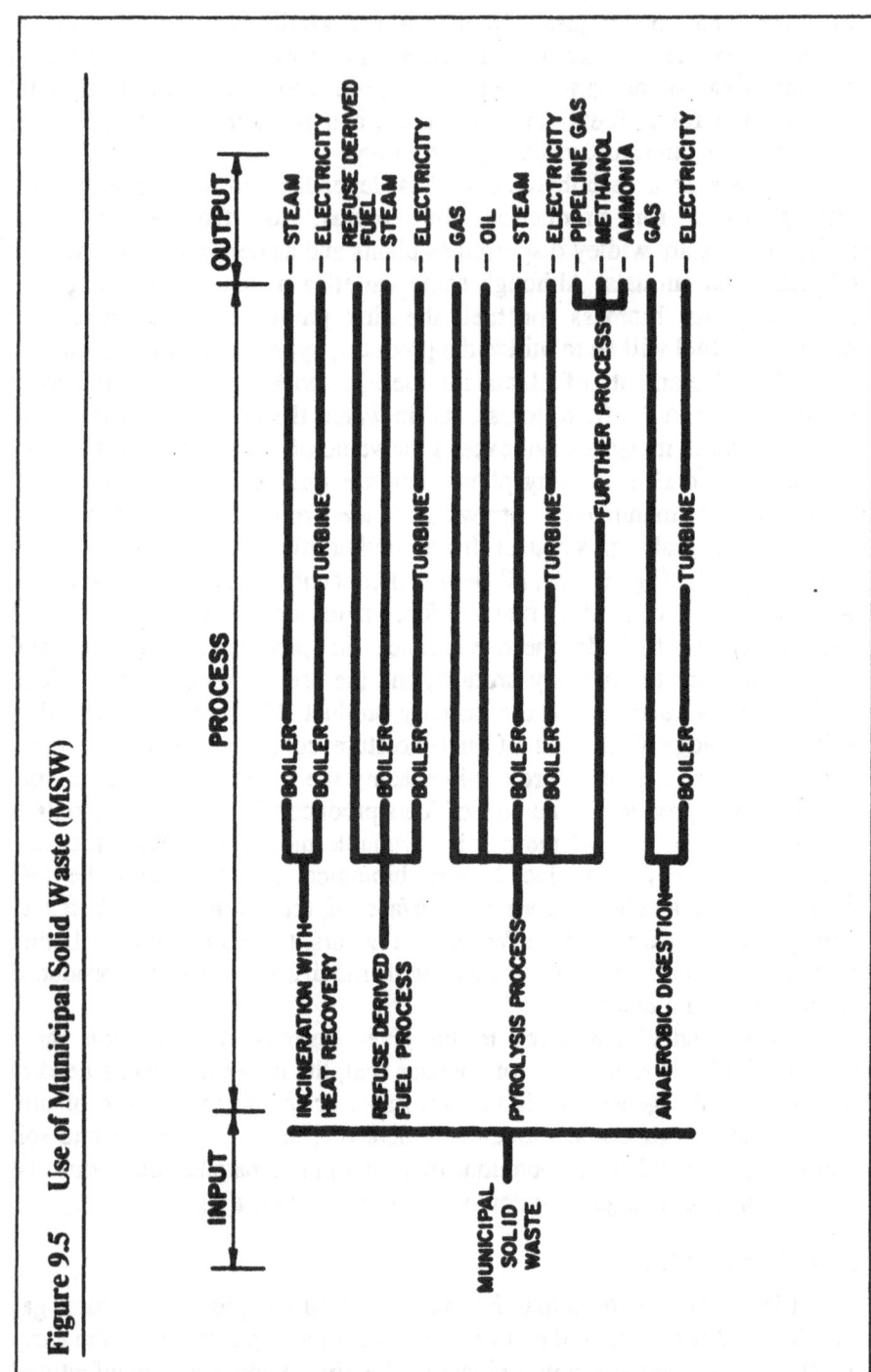

286

gas, consisted of vaporizing coal by retort, resulting in vaporization of about 20 percent of the coal to gas. The manufactuared gas was medium-Btu content, 550 Btu/cu ft (20 MJ/cu m). Later in the 19th century, another manufactured gas known as water gas developed. Water gas was made by moving steam over hot coke, producing carbon monoxide and hydrogen. The gas was low-Btu, 275 Btu per cu ft (10 MJ/cu m). These products could come back into the picture if the availability of natural gas would greatly decrease, a pipeline distribution system were in place, and coal resources were available. If coal or precursors to coal were available at a relatively low cost, a replacement gas could be produced, probably at a lower cost than using biomass. Although environmental problems exist for using biomass as a resource for making gas, the environmental problems for making gas from coal are generally considered to be greater -- potentially higher sulfur content, possibly higher radiation level, need to dispose of ash, and the challenge to prevent air pollution.

Use of Pipeline Systems

The availability of natural gas in many developing areas of the world provides a basis of energy supply for further development. That energy can be part of the driving force for development of chemical feedstocks, production of electric power, and for heating processes. To provide for use at many different locations on an economical basis, a pipeline is needed. Those people responsible for money to be invested in development, particularly if the quantity of natural gas and the time of availability are not known, could well insist on long-range plans on the use of the natural gas system over a long period of time -- say 30 yr for a 30 yr loan. What are the options and prospects?

The prospects of finding new sources of natural gas in the areas presently identified, particularly at deeper locations and in areas where coal and petroleum are available, additional natural gas may be found. These trends have occurred in several areas of the world.

If the country has a port or access to ocean transportation, liquid natural gas could be purchased and fed into the pipeline system with no noticeable impact on the users. Transfer stations are required at the entrance to the pipeline as the liquid changes to a gas, leaving the tankers transporting the fuel.

The natural gas pipeline can be used for other fuels that duplicate natural gas (high methane) or for manufactured gas that substitutes for natural gas, the major components of which are mainly carbon monoxide, hydrogen, and methane.

287

If coal and petroleum resources are available, these can serve as raw materials to manufacture gases that substitute for natural gas. Usually, petroleum should be reserved for lubrication or liquid fuels for vehicles, rather than gaseous products for fuel.

In areas with appropriate soil conditions, acceptable slope of land, and farmable weather, biomass can be used as a resource for making methane or for making substitute gas for natural gas. Sewage and waste products from food and wood processing plants can be converted to methane. These resources, accumulated after processing, are concentrated in one or a limited number of locations and provide a ready-made resource. These resources are often of high moisture content requiring part of the energy to dry. As a fuel source, biomass grown in the fields, forests, or in the lakes and oceans, needs to be grown, cultivated, harvested, transported, and processed. These are all expensive operations, probably more costly than the value of the product itself. Eventually, after other more readily available resources become limited and more expensive, biomass, which will also become more expensive to produce, provides a viable alternative as an energy source. Generally, the closer the material resource, in a process state, is to methane, the more economical and less expensive, comparatively, is the conversion.

Thus, natural gas is presently the least costly source for methane. On the other hand, greater drilling depths, along with higher costs of exploration, will increase the cost of natural gas.

Concerns for the environment are paramount in today's world. By using natural gas as a fuel, there is little impact at the source. Oil wells with methane no longer flare gas into the environment, and measures should be taken to minimize leakage into the atmosphere. At the combustion end of the system there is a concern for the effect of the products of combustion, particularly for gas with sulfur compounds. However, the impact is less with natural gas than with other fossil fuels for the same amount of heat produced. When fuels are manufactured, major concern exists for the impact at the manufacturing end, adding a source for environmental impact. And for all the advantages of using the biological cycle to provide methane, all be it similar to the natural production of methane, environmental impacts need to be carefully considered.

Additional costs are involved when a pipeline originally designed for natural gas (methane) is used for other gases (carbon monoxide, hydrogen, mixtures which include some methane). The basic pipeline can be used, just as early pipelines were converted for use form non-methane to methane gas. As examples of variation of fuels are the thermal content of the gas (if lower thermal content, larger pipe or higher pressure needed), removal of impurities (such as sulfide compounds), safety (range of explosibility in air varies greatly), and toxic or lethal aspects (carbon monoxide is highly poisonous).

Summary

Natural gas has been discovered in many developing areas of the world. Natural gas provides an economical energy source which can be used for many applications from direct heating, in the simplest, to conversion for chemical feedstocks. To provide the greatest potential for development, the natural gas needs to be available to numerous customers, not a few large customers. A pipeline network connecting sources to users is commonly the pattern.

As natural gas supplies become depleted, the pipeline can be used to distribute gas of various composition, synthetic natural gas or manufactured gas, so that the developing country, at first dependent on natural gas, can be confident of continuing its growth and providing employment for its citizens, on the next best alternative available at the time. In many areas of the world the reserves of natural gas are of sufficient quantity to justify the pipeline distribution system without serious consideration of alternative gaseous fuel resources. A national or international pipeline provides the network by which there is considerable interchangeability of fuel, making possible the best utilization of resources (hydrocarbon or carbohydrate), continued growth of industries and employment of people, with minimum impact on the above-ground transportation (trucks, trains, and boats), while minimizing environmental impacts, particularly at user sites.

REFERENCES AND NOTES

1. Ahlberg, K., *AGA Gas Handbook*, Almquist and Wiksell International, Stockholm, Sweden, 1985.

2. Bridgwater, A. V. *Thermochemical Processing of Biomass.* Butterworths, London and Boston, Mass.,1984.

3. Hall, Carl W. *Biomass as an Alternative Fuel.* Government Institutes, Inc., Rockville, Md., 1981.

4. Hall, Carl W., and Hinman, George W., *Dictionary of Energy*, Marcel Dekker, Inc., New York, N.Y., 1983.

5. Kirk-Othmer. *Encyclopedia of Chemical Technology*, Wiley Publishing Co., New York, N.Y., 1978-1974.

6. U.S. General Accounting Office, *Conversion of Urban Waste to Energy*, Washington, D.C., 1979.

Appendix A

The Importance of Gas Technology to Future Natural Gas Supply and Demand

Much of what has been written and spoken about the gas industry focuses on the economic health and well-being of the industry. The shift to a more competitive environment in the US and Europe, mergers and acquisitions, etc. all are being played out on center stage. But in the wings there has been a more subtle, but potentially even more important evolution: the growth of modern gas technology.

The technological changes in the past 20 years have been truly impressive and dramatic. They have ranged from developing more efficient sources of gas supply, to better ways to transmit gas, and more efficient ways to use it at the burner tip. Significant as these developments have been, this is only the beginning. There are more sweeping developments on the drawing boards today that will minimize the cost of supplying natural gas into the 21st century while maximizing the value of the energy provided. The purpose of this Appendix is to chronicle the growth of this technology and point some direction for its future.

Without adapting to changed environments, there can be no growth. This is true for companies as well as organisms. The future success and growth of the gas industry will depend on its ability to adapt to change. In the past, gas technology developments were responses to problems. For example, when there has been a question about gas supply, gas technology moved in to prove those concerns were false. However, if technology is to lead the gas industry to a greater and more efficient future, then technology cannot be simply a problem solver -- it has to be a stimulus for new markets and provide the basis for confidence that they can be efficiently served. There are institutions in the gas industry -- such as the Gas Research Institute, the Institute of Gas Technology and the Gas Appliance Technology Center at the American Gas Association (A.G.A.) Laboratories, just to name those in the United States -- that recognize the need for linking technology to commercial development. These private sector technology initiatives are now making their mark.

There remains a continuing need for long-range, more fundamental research that looks at gas and new sources of gas from the viewpoint of fundamental energy options. This area of technology development will still need strong and focused funding support from the governments. Taken together, private sector and public sector funding should keep gas technology in front of, and leading, the gas industry into the more competitive environment of the future.

Good Ideas
Precede Good Technology

In the past, technology development responded to market requirements. For example, over the last 20 years, supply has been a major issue for the industry. Therefore, on the supply side, a greater expenditure of research and development (R&D) resources has provided a number of direct benefits, such as: deep drilling onshore, deep water drilling offshore, better seismic capability, and greater knowledge of tight gas formations. The successes in these areas over the last two decades have helped to redefine the gas supply outlook from "scarce" to "plentiful."

In the future, research should anticipate market needs. On the demand side, many products of new technology are just becoming available today. New gas products include high efficiency heating and cooling equipment, improved gas turbines, industrial burners and furnaces and natural gas-fueled vehicles. Despite these impressive accomplishments, the demand side hasn't yet seen the dramatic R&D breakthroughs that have characterized supply R&D.

History

The resource base is dynamic, changing as new technology opens new resource possibilities. This process is often difficult to see in the short-run, but it has major implications for R&D planning. In the past, energy planners believed that natural gas resources were playing out over time and directed R&D funding towards electricity, primarily nuclear energy. History has proven that technology developments can increase gas supply, and sources that were not considered to be recoverable 20 years ago are now classified in the conventional resource base.

The ability to drill for gas in offshore waters is one example of this process. The record in 1965 was 632 feet of water; by 1984, Shell had set a new record in 6,448 feet of water. The theoretical limit today is around 10,000 feet. According to the estimates of the Potential Gas Committee (US), the increases in water depth capability from 1965 to the late 1970's increased our offshore recoverable resource base by 50 percent. At this rate this technology growth has major implications for future additions to the recoverable resource base.

In the area of drilling capability onshore, similar statistics are available. In 1938 the record depth for a well was 15,004 feet. Today the record depth is 31,441 feet -- more than double the depth. From a recoverable resource base view in the lower 48 states, the increase in depth capability from 15,000 to 30,000 feet increased the recoverable gas resources by 86 percent.

This technological growth was, in effect, the result of the petroleum industry trying to develop oil and gas resources. If governments, together with institutes and private sector gas companies, developed the needed technology today, similar, if not greater, technological advances tailored to develop natural gas resources could be made on an accelerated timetable. There is tremendous potential for increased gas supply from sources that are *not now in the recoverable resource base*. Some of these sources are being developed today -- some are clearly long-term investments. But as deep water gas and deep well gas are being added to recoverable resources, US energy planners should be looking ahead to the next gas resource potential.

Short-Term Supply Potential

The payoff from past technology developments is clear. Today research is being directed towards achieving commercial gas production from unconventional resources.

To make unconventional resources economically and technically producible requires advances in recovery techniques. A key area for gas research is tight sands -- gas-bearing formations characterized by low permeability. Currently, to take the US as an example, 172 Tcf of tight sands gas is considered recoverable, and therefore is included in the conventional resource base. This 172 Tcf transferred over the past 5 years from unconventional to conventional is the result of technical improvements. However, estimates of the in-place resource are as high as 900 Tcf.

Another major gas supply potential is the Devonian shales, estimated to contain more than 1,800 Tcf of gas in-place. To increase production from tight sands and Devonian shales, more research must be conducted to develop advanced stimulation techniques to fracture tight rock formations that restrict gas flow. Research must also focus on field testing of production techniques and establishing a data base for analytical purposes. The existence and location of these gas deposits are fairly well defined and the development of the right technology could yield significant results in the near term.

Incremental methane also can be obtained from the co-production of brine and gas. Many water-driven gas wells have been abandoned when water encroachment floods the well. Research into techniques to produce gas in reservoirs where gas wells have "watered out" could increase the amount of gas amenable to co-production.

293

Intermediate Supply Potential

Several gas supply sources are not now included in the remaining recoverable resource base estimates. Research could contribute positive results in the intermediate- to long-term time frame. Coal seam methane is one such supply option. In the US alone, this total resource is estimated to be as high as 800 Tcf. In recent years there has been significant progress in the development of techniques which optimize the production of methane from coal seams. Although conventional gas production techniques generally have been unable to produce this gas on a commercial basis there have been several commercial operations underway. Further improvements will be needed before most of the resource is recoverable.

In addition to naturally occurring gas, gas can be produced synthetically from either coal, municipal and agricultural wastes or biomass. There are over 30 active waste gas projects and one major high-Btu coal gasification project underway in the US. These projects have, in general, proven to be more technically and economically successful compared to other synthetic gas projects. With continued R&D support, these sources could make a significant contribution to gas supplies.

Another potential for natural gas lies in geopressured areas. One of the largest geopressured areas underlies a portion of the Texas and Louisiana Gulf Coast. Estimates of the total resource in-place range up to 100,000 Tcf. Even the most conservative estimates of the amounts of gas which might ultimately be recovered exceed 25 Tcf and, depending upon the degree of R&D success, geopressured zones provide a 200-year plus source of domestic energy potential.

Interest in the geopressured gas research at the US Department of Energy (DOE) has been on the decline. The focus of the work should be to emphasize the research benefits of the geopressured natural gas under an entirely different set of assumptions than the traditional thermal approach taken by DOE. The "controlled blow out" theory advocated by Dr. Paul H. Jones of Louisiana State University is contrary to the traditional thermal approach and would focus on standard reservoir engineering principles to maximize gas production and minimize brine production. This approach is now gaining support at the World Bank for application in Asia. This vast natural gas potential has not been adequately explored and current programs are not addressing the problem on a realistic scale.

Long-Term Supply Options

Long-term gas supply research efforts are those that have little likelihood of contributing to energy needs this century. These are high risk, high payoff technologies, which if proven, have the potential of making natural gas a virtually inexhaustible energy source. Research efforts in gas hydrates and abiogenic gas have essentially been ignored in the US. Worldwide, nations that have only recently acknowledged the benefits of natural gas have funded significant research efforts in these areas -- specifically the Soviet Union in gas hydrates and the Soviet Union and Sweden in abiogenic gas.

Gas hydrates are solid, ice-like compounds in which gas molecules are entrapped and bound to water molecules. Recently, it has been recognized that methane hydrates occur naturally in some permafrost zones and ocean sediments. While most of the research on gas hydrates in sediments has been conducted in the Soviet Union, drilling in the North American Arctic and the offshore Atlantic have reportedly confirmed their existence. Typical estimates of the total world resource of gas in hydrates range to over 270 million Tcf. This entire potential resource is totally excluded from current recoverable resource estimates.

There are a number of theories which suggest the existence of large volumes of methane deep within the earth's crust and upper mantle. One theory concludes that massive amounts of abiogenic methane, that is, methane whose origin is of a non-biological nature, were derived from a primordial supply of hydrocarbons trapped during the earth's formation. There is now sufficient evidence available to conclude that at least small amounts of the natural gas being produced today are the product of outgassing of methane from deep within the earth. Many traditional geologists now believe that both biological and abiogenic theories of gas formation can coexist. Scientific evaluation of this theory could lead to a new and more complete understanding of methane production -- opening up new drilling prospects that formerly were overlooked.

DOE programs in hydrates and deep source gas are promising, but too small to produce the desired results. These resources are geographically dispersed, pose complex production challenges, and could be extraordinarily expensive to develop. The research program must be broad enough to respond to those questions and make these resources economically producible early in the next century.

The Role of Geosciences

Gas supply research should be considered as a major component of energy technology development. The outdated belief that the natural gas resource base limits supply potential not only incorrect for practical purposes, it is also self-defeating. The gas resources in-place in the US are potentially several thousand times the current consumption level.

Basic geosciences research is essential to insure proper development of natural gas resources. This discipline -- understanding fundamental geology -- traverses all gas resources. As improbable as it seems today, a technological breakthrough tomorrow from any one of the exotic sources -- geopressured zones, gas hydrates, abiogenic methane -- could lead to a truly inexhaustible supply of methane. Technology improvements have already resulted in better drilling economics. Future advances in gas supply technology should provide a stimulus for new markets. Basic research in the geosciences is thus essential for the future. But with a long-term program of scientific evaluation and technological progress, it is within the realm of possibility that natural gas could turn out to be our largest energy resource base.

New Gas Utilization Technologies

Advances in technology for natural gas at the point of end use already have increased the efficiency of gas appliances and processes. In each market sector, cost-effective equipment is improving gas service with significant cost savings.

In the US, the A.G.A. Laboratories have been responsible for much of the research in residential and commercial gas appliances and space-conditioning equipment. In the residential sector, furnace efficiencies are in the 80-97 percent range, compared with 55 to 65 percent efficiency prior to 1980. Water heater improvements have increased thermal recovery efficiencies from less than 70 percent to 90 percent during the same period. High-efficiency room heaters are now undergoing field tests.

Efficient, cost-effective commercial equipment, including cogeneration systems and gas cooling systems currently on the market, will help expand the role of gas in commercial space conditioning. High-efficiency commercial water heaters and cooking equipment also have been introduced to the cost-conscious business and restaurant sector. The pulse combustion principle, successfully applied to gas furnaces and boilers, has been expanded to commercial cooking.

The next generation of gas equipment will need advances in technology to improve the ability of a single piece of equipment to provide multiple energy services, with significant cost savings and efficiency gains. Natural gas cooling is one example of how technology first limited and then substantially improved the market potential for this end use. Gas air-conditioning was promoted heavily in the 1950's and 1960's with some success. However, in the 1970's, while the gas industry encountered problems with regulation-induced supply shortfalls, the electric industry was wining most of the growing air-conditioning market, first with conventional air-conditioners and then with heat pumps. Gas-fired equipment had a difficult time competing, and there was little support for gas cooling R&D.

With the improvement in gas supplies and prices, market conditions have changed over the last few years. A new generation of gas cooling technology for commercial buildings has been introduced to the US market by Japanese manufacturers. Double-effect direct gas-fired absorption chillers/heaters are being manufactured by three Japanese firms. This equipment has higher efficiencies than the earlier gas air-conditioners. Research and development in the United States is now underway for gas cooling and combined heating and cooling technologies, and gas cooling now has real growth potential. But that growth potential is coming from foreign R&D.

Natural Gas:
The Competitive Edge in Industry

On-going R&D programs in natural gas utilization are making significant contributions to developing the most effective technology to convert fuel into useful energy. For example, natural gas-fired heat pumps now under development could save one half of the gas used in a conventional furnace and one-quarter of the gas used in a high-efficiency furnace.

The many attributes of gas, such as convenience, ease of handling, stable flame characteristics and chemical composition, make gas suitable for a wide variety of industrial applications. Heavy industry is a sector where US firms face increasing international competition. Natural gas can give the industry an edge by providing a lower-cost energy source that is efficient and has low capital requirements.

New gas technologies are already improving the performance of heavy industry. Oxygen/gas burners, for use with electric arc steel-making furnaces, can increase furnace production output by 15 to 30 percent while reducing energy costs by 10 percent or more. Regenerative gas burner systems may be retrofitted to existing high-temperature gas furnaces, thereby providing a 40-50 percent efficiency improvement. Mechanical vapor recompression systems allow the recovery of energy from waste steam by compressing low pressure steam and returning it to a usable pressure.

With additional R&D, large gas-fired cogeneration systems and heat pumps will come into the marketplace. In the long term, R&D must focus on materials and components and the interaction of gasification catalysts with carbon/steam and carbon/hydrogen reactions. Advances in catalysis concepts, for example, could open natural gas to all petrochemical applications.

Gas-fired equipment used for electric generation represents a new, higher efficiency and environmentally benign means of producing electricity. Research in gas-fired cogeneration could produce overall efficiencies of 70 to 80 percent compared with the average of conventional primary electricity generation efficiency of 33 percent. Combined cycle gas turbines for central station electric power generation are now 46 to 50 percent efficient, again compared to 33 percent for conventional electric power generation.

New gas combustion technologies can also reduce industrial and power plant emissions. The select use of gas with coal can significantly reduce emissions of sulfur dioxide (SO_2) and nitrogen oxides (NO_x). Retrofitting a boiler for gas-enhanced combustion, where natural gas is used as a "reburn" fuel inside the boiler, can be 50 percent more effective than using coal reburn.

Long term basic research can introduce new processes to change gas into usable energy -- electrochemical processes and thermodynamics. The gas-fired fuel cell represents the first generation of this type of application.

Natural Gas Vehicles

In the transportation sector, natural gas vehicles represent a market opportunity in a number of cities. They also provide an opportunity for cities to reduce their air pollution problems. Over 500,000 natural gas vehicles are on the road worldwide, with some 300,000 more planned by the early 1990's.

Further research can optimize system designs for natural gas-powered vehicles, and find novel concepts for gas compression and storage. With these advances, natural gas could become more universally accepted as a transportation fuel.

Efficient and Safe
Transmission and Distribution

The reliability and efficiency of natural gas have been responsible for the steady and considerable growth in its usage over the years. In response to the changing needs of the industry, gas pipeline companies are planning a number of ways to expand and strengthen their service in coming years. Ensuring that the natural gas transmission system remains one of the safest major transportation systems in the world is an overriding goal. The industry also must continue to improve efficiency and cost-effectiveness in gas transmission and related service operations.

298

The use of plastic materials is a rapidly expanding technology throughout the gas industry. Plastic pipe has many advantages. It is less costly and easier to handle and transport, not subject to corrosion and can be installed without welders. Other developments include service truck mounted computer terminals, and the remote and automatic reading of meters, including instant transmission and readout at a central office. In addition, energy generated by the pressure reduction of gas upon delivery at the city-gate can even be recovered to generate electricity for the gas utility.

For the future, pipeline R&D conducted by the A.G.A. Pipeline Research Committee will focus on cost-effective techniques for operating, constructing and maintaining the pipeline system. To increase understanding of how and why pipeline service failures occur, to improve gas storage options and to design on-site industrial storage options will require R&D in the materials and engineering areas. The environmental impacts of emissions from pipeline facilities although very minimal, will continue to be evaluated. Advanced materials research will provide new pipeline products in the future, and improved compressor efficiency can reduce transport costs. For example, several transmission companies are currently using microprocessor controls to load compressors, with reductions of 5 to 10 percent in fuel use. In the long term, basic research in the principles of gas flow and measurement can reduce system costs significantly.

Conclusion

Technological advancements in natural gas supply, transmission and utilization have been signficant. But what was a trickle yesterday is turning into a wave of new techniques and equipment. And technology requirements in the future will continue to need innovative, even exotic R&D concepts and applications. The gas industry cannot afford to undertake this long-term, high-risk commitment by itself, and the federal government cannot afford to ignore the future potential for natural gas. Therefore, a joint industry/federal effort is necessary. As the industry has adapted its operations and marketing strategies to compete in a decontrolled environment, it must also continue to adapt new technology to productive applications for the future.

Gas research is changing the world energy outlook. New gas utilization technologies such as gas cooling are displacing peak electric requirements, while cogeneration and combined-cycle equipment are providing more efficient new electric capacity. Novel gas technologies such as high-efficiency furnaces, new industrial applications and natural gas vehicles are also backing out imported oil. Gas supply research is expanding supply alternatives to satisfy these new demands, while new transmission and distribution technologies are enhancing the effectiveness of the transportation system. And so, as most policy planners focus on the structural changes in energy industries and the regulatory arena, the gas technologists are quietly stealing the show.

Appendix B

Further Description of
Natural Gas Resources and
Gas Exploration and Production Technology

The purpose of this Appendix is to describe the natural gas resource base and the role of technological improvements in bringing the resource to market.

Nonconventional Sources of Methane

There are a number of resources from which future gas supplies will be produced as new technologies are developed, existing technologies are refined, and the economics for natural gas supply change. The gas industry has actively supported research and development programs designed to bring about the technological improvements necessary to foster production of these sources.

Included in the nonconventional category are methane from biomass and waste, in-situ coal gasification, peat gasification, oil shale gasification (each of which involves a process whereby methane is produced from the conversion of solids) and methane hydrates, (which involves the production of gas from naturally occurring deposits). Also included in the nonconventional category is methane from landfills, production of which is already taking place on a limited scale.

The sheer size of the resource base associated with each of the nonconventional sources continues to encourage research related activity for developing the technologies needed for methane recovery from these sources. However, with the strong near-term outlook for gas supply from conventional and other supplemental sources, progress in bringing forth methane production from the nonconventional sources is expected to be slow; thus, the focus on future production has shifted to the post-2010 period.

Biomass and Waste. The term "biomass," as used in this appendix, refers to plant matter grown and harvested specifically for the production of energy. A variety of crops are being examined for this purpose including various grasses, plants, and trees in onshore locations, and kelp and other seaweeds in the oceans. "Waste" describes the organic

301

byproducts, both animal and vegetable, which result from activities not specifically oriented towards energy production.

The production of methane from biomass and waste involves the use of processes which, through either anaerobic digestion or thermochemical conversion, converts solid (or liquid) organic material to a gaseous form which must then be processed to produce a high-Btu pipeline quality product. Numerous above-surface processes have been developed to convert biomass and waste, however with few exceptions they have yet to demonstrate commercial applicability.

In certain geographic areas the potential supply from these sources could be quite significant in terms of local energy needs. However, the widespread availability of conventionally produced resources has in recent years served to delay the development of biomass energy technologies. Nevertheless, there is industry-wide recognition of the tremendous potential which the biomass and waste resource represents, thus limited research and development programs needed to bring about production of this supply source continue.

Since the biomass and waste resources are renewable there are no limits to the amount of energy which can ultimately be produced from these sources. However, annual production levels will be limited and will depend on:

- the development of conversion technology which optimizes gas yield and reduces the need for large capital investment.

- the establishment of controlled cultivation systems which will produce commercially viable crop yields.

- the development of methods to process residuals into commercially valuable by-products.

- further development of harvesting, transportation and handling systems.

- methods of protection from diseases, insects and weather being established for biomass crops.

- changes in energy markets relative to the economics of energy production from biomass.

Environmental considerations, such as the need to dispose of digestor residuals or the impact of large-scale marine biomass farms, must also be further evaluated.

Methane from Landfills. Among the techniques which have been developed to recover the energy in waste is the production of methane

302

from landfills. Since its inception in the early 1970s, many projects have been developed producing medium and high Btu gas. Medium Btu gas is typically used in a local industrial application whereas high Btu gas is pipeline quality and is generally directly linked to nearby distribution systems and provides a secure local supply of gas for the local utility companies.

Landfill gas (LFG) is generated by the bacterial decomposition of organic materials contained in municipal solid waste (MSW). The rate of production and composition of the gas will vary depending primarily on the waste loading rate, the age of the landfill, the composition of the MSW and the moisture level.

Once deposition of the MSW has commenced and methane formation has started, the gas must be controlled and collected. Collection systems typically consist of a series of wells drilled into the landfill which are connected by a gathering system.

Extraction of the LFG is achieved by applying a vacuum to the system which withdraws the gas from the wells. It is important that air is not drawn into the landfill during this process as it will destroy the activity of the methanogenic bacteria.

The collection system transports raw LFG to a nearby treatment facility for upgrading. The amount of upgrading depends on the intended application for the gas. For high-Btu gas, most of the carbon dioxide contained in raw landfill gas (about 50 percent by volume), as well as all other compounds, except methane must be removed. For both high- and medium-Btu applications, the gas must be dehydrated and compressed before injection into a pipeline. A distinct advantage of producing high-Btu gas is its ability to be transported through the existing distribution network thereby adding to peaking capacity.

Resulting from the ability to recover methane from landfills are a number of additional benefits. Among the most important are:

• A reduced hazard from uncontrolled accumulations of methane and capturing a usuable resource.

• The reduction of undesirable odors.

• Environmental advantages due to the reduction of reactive organic pollution problems and the destruction of vegetation which can result from gases migrating to the surface.

• Increased local energy supply security. LFG recovery provides a source of energy supply very near the eventual point of consumption. This increases the security of energy supply to the local community.

- Economic benefits to the local community from payment of taxes and royalties to the owners of the landfill (often city governments).

- Conservation of non-renewable fuels.

There are usually no severe technical limitations with respect to extracting and/or delivering landfill methane to meet individual customer needs and efforts to further improve the technology are continuing. The limitations of this market are economic. The decline in the real cost of oil and gas, combined with an inability to substantially reduce the cost of constructing and owning such facilities, no longer make them competitive with other incremental sources of energy. Thus, it is unlikely that additional plants will be built in the near term except for environmental reasons whereby the uncontrolled generation of methane from these sources present a safety hazard or are determined to be a significant contributor to free methane in the upper atmosphere. In the longer term, economics may again be favorable. Existing plants with long term contracts are likely to remain operational.

In-Situ Coal Gasification. Much of the coal resource is located in deposits which are not minable with conventional mining techniques. However, recovery of the energy from these deposits may be accomplished through in situ coal gasification, that is, conversion of the coal to gas underground in the deposit where the coal is found. Gasification in surface facilities to produce high or medium-Btu gas has been studied in great detail and thus much is known about the gasification process. In-situ coal gasification relies on essentially the same chemical processes as surface gasification; however, because the coal is not mined, in-situ gasfication does not require the relatively expensive mining and coal handling systems required in a surface facility.

In-situ coal gasification has the potential to significantly expand economic extraction from coal seams which are too deep, too steep or too thick for conventional mining methods. In underground coal gasification, burning is initiated at a primary bore hole, and the product gas flows through a physical linkage to another bore hole for recovery.

Although gasification of coal in-place simplifies some aspects of the conversion process, a variety of technical questions must be answered before significant amounts of gas can be economically produced. In addition, there are environmental uncertainties associated with underground coal gasification: subsidence causing aquifer disruption, water quality effects caused by contamination of ground water with organic and inorganic materials, surface disruption, water usage, possible health and safety problems, and waste and atmospheric emissions.

Peat Gasification. Peat, geologically the youngest form of coal, consists of partially decomposed vegetation which has been deposited in bogs at the land surface where moderate to high rainfall and partial isolation from local drainage systems create anaerobic conditions. As a result, peat occurs at the ground surface with little or no overburden material.

The peat resource base is less well defined than other fossil fuel resources, although enough is known about certain areas to justify planning for future gas production.

In its natural state, peat may contain over 90 percent water by weight. In order to use peat as a fuel or for gasification, the bulk of this moisture must be removed. In Europe and the USSR, where peat is commercially used as fuel in power plants as large as 730 megawatts, peat harvesting technology that incorporates air-drying in the drained bogs is widely used. Such techniques, however, are not very attractive in many peat-rich regions in the United States. Development of novel and economical techniques for dewatering peat is critical to its use as a resource for SNG production.

The gasification of peat to produce low-Btu gas has a history that goes back more than a hundred years. Peat gasification has been conducted on large scales in many different types of gasifiers. Commercial processes for the production of low-Btu gas from peat are offered by a number of manufacturers. It was only in 1978 that work was directed toward production of SNG from peat. Most of this work has been conducted in the US at the Institute of Gas Technology (IGT) from laboratory through pilot-plant scale. This research has shown that, compared to coal, peat is more easily gasified and produces more methane, making peat a better resource than coal, on the basis of chemistry, for SNG production.

In late 1985, peat from Finland was successfully gasified by IGT at a rate of 1 ton/day in its U-Gas pilot plant for the production of medium-Btu gas. The medium-Btu gas produced can be upgraded by state-of-the-art processing to SNG.

A commercial-scale plant is nearing completion in Finland for converting peat to synthesis gas for ammonia production. The plant is capable of processing 650 tons/day of peat and uses the High-Temperature Winkler process for gasification.

Oil Shale Gasification. Oil shale is a fine-grained sedimentary rock containing an organic material known as kerogen. Upon heating (retorting), the kerogen in shale decomposes to yield oil, gas, and residual carbon. The research conducted at the Institute of Gas Technology (IGT) has shown that retorting of shales in the presence of hydrogen at elevated pressures increases the yields of oil and gas. The improvement achieved by

305

this approach is much more noticeable for those shales where carbon recovery of up to 2.5 times over that by conventional retorting can be attained. The raw shale oil is more like conventional petroleum than coal-derived oil, and can be further processed into liquid and gaseous fuels.

In late 1983, pilot-plant scale (8 ton/hr) shale gasification tests were conducted in Sweden by Aktiebolaget Svensk Alunskifferutveckling (ASA) for the production of medium-Btu gas using a slagging gasifier. This technology should be applicable for processing other oil shales. Research is also being conducted to reduce the mineral matter content of oil shales prior to retorting. Technology developed to beneficiate coals is being applied to separate kerogen from the mineral matter in oil shales. Beneficiation of oil shale very significantly reduces the cost of shale conversion by IGT's hydroretorting process. Research programs are under way to produce medium-Btu gas from oil shale without using hydrogen and by in-situ gasification techniques using indirect heating.

A larger-scale pilot plant and demonstration plant operations are needed to verify the technical and commercial feasibility of converting oil shale directly to SNG.

Methane Hydrates. Methane hydrates are solid, ice-like compounds in which, under certain conditions of temperature and pressure, gas molecules are entrapped and bound with water molecules in a crystalline structure. It has been theorized that because conditions favorable to gas hydrate formation exist on a large portion of the earth's surface, significant accumulation of methane in gas hydrates may exist in deep ocean deposits and in permafrost areas onshore. These accumulations would include methane trapped in the gas hydrates themselves or possibly large quantities of gas trapped in the sediments below the impermeable gas hydrate layers.

Estimates of the resource for methane from hydrates are generally inferred based on theoretical knowledge about the conditions favorable for hydrate formation and the confirmed existence of hydrates in a number of locations. The limited amount of information results in estimates which are highly speculative, however all estimates indicate an enormous resource -- possibly as high as 270 million Tcf worldwide.

Currently, there is no established production technology for recovering methane from hydrates although a number of concepts have been proposed. In general, these involve pressure reduction, temperature increase, and/or the addition of chemical agents to dissociate or "melt" the solid hydrate. An important consideration in the evaluation of production technology is the net energy balance, that is, whether the energy recovered from the methane in the hydrate exceeds the energy input required by the production system. Given the low permeability of hydrate-cemented

306

sediments, the injection of fluids to "melt" the hydrates would be difficult. In addition, production systems may be required to handle the large amounts of water which would be released as hydrates are dissolved. Since much of the resource is located offshore, the problems unique to deep water drilling and production would also be a factor.

New Resource Development -
Coal Seam Methane

One example of a new category of natural gas resources which is being added to the estimates of recoverable resources is methane from coal seams. This resource which was only sporadically produced and then only for low or medium Btu applications twenty years ago is now a prime target for development drilling in the US. The emergence of this resource as a practical, productive source of gas is apparent as the annual US production grows from 27 Bcf in 1987 to 50 Bcf in 1988 possibly to 800 Bcf by 2010.

Coalbed Methane Background. Methane trapped in coalbeds has long been recognized as a substantial unconventional energy resource. Estimates of gas in place and the recoverable resource vary widely. In many basins, resource estimates have historically included only shallow coal seams leaving potentially methane productive deeper coals unaccounted for. Also, from region to region coals vary considerably in methane saturation per ton. Whatever the difficulties in resource estimation some facts stand out. Production of methane from coalbeds exceeded 27 Bcf in the US in 1987. That figure will likely exceed 50 Bcf for the year 1988 according to the Gas Research Institute. While still below one percent of total U.S. gas production, several companies are strongly committed to the growing development of coalbed methane reserves.

Origin and Occurrence. Methane is a normal product of the process that converts accumulations of plant matter into coal. Some methane is produced biologically by organisms during the fermentation of wet, buried plant material. Additional methane is generated when coals are buried at depth and subjected to thermal maturation. Heat and pressure "cook" the source material, but, unlike conventional sand reservoirs where gas is generally accumulated after migration, coal acts as its own reservoir, also. At a microscopic level coal is sufficiently porous to hold substantial quantities of methane molecules. However, fractures or cleats (regular fracture patterns in coal) are required to create permeability so that gas can flow through the coal.

Geologic conditions at the time of deposition and resulting burial history influence greatly the type of coal developed and the amount of methane produced and trapped per ton of coal resource. Pennsylvanian time (ending 280 million years ago) was a time of favorable coal deposition in the eastern United States. Coals of the western United States were deposited in more recent Cretaceous and Paleocene time (from 135 to 58 million years ago).

Gas Production Characteristics. Techniques for completing coalbed methane wells vary from operator to operator. But, certain characteristics of coalbed methane production persist. Methane in coalbeds is normally associated with water. Wells penetrating coalbeds fill first with water. Often bottomhole pressures are insufficient to produce gas through the water column. Water is then lifted with the aid of a surface pump and methane is produced with water (for disposal). Over time the coals tend to dewater; water production declines and stablizes and gas production rises. This produces a "reverse" production decline curve from that of traditional reservoirs. Though gas production eventually declines, many coalbed methane wells produce at steady rates for many years. Well life has approached 30 years in some coalbed methane producers.

Low steady production rates may create problems as well. Reservoir pressures may require that compression facilities be set to produce gas. Water disposal can also add expense to a well completion. Technology can help the producibility of some coalbeds. Horizontal drilling technqiues can enhance wellbore contact with the principal permeability mechanisms (cleats). Such a horizontal development pattern might reduce the number of wells necessary to drain a given reservoir as opposed to a normal pattern of vertical wellbores.

Appendix C

Terms and Conversion Factors

The following are some frequently used units of measure and natural gas equivalents in British thermal units (Btu). Read "=" as "contains" between volumes and energy units.

The national average Btu content per cubic foot is used.

1 Cubic Foot (cf) = 0.02832 Cubic Meters = 1,031 Btu[*]
1 Cubic Meter = 35.315 Cubic Feet
1 Therm = 100,000 Btu = 25,200 Kilocalories = 100 Cubic Feet

1 Mcf = 1,000 Cubic Feet = 28.32 Cubic Meters = 1,031,000 Btu[*]
1 MMcf = 1,000,000 Cubic Feet = 1,031,000,000 Btu
1 Bcf = 1,000,000,000 Cubic Feet
1 Tcf = 1,000,000,000,000 Cubic Feet
1 GigaJoule = 1,000,000,000 Joules; 1,000,000 Joules = 947.8 Btu
1 quadrillion Btu (quad) = .9709 Tcf[*]

For comparative purposes, Btu contents of various energy sources other than natural gas are shown below.[*]

Fuel Oil Distillate (barrel)	5,825,000 Btu
Residual Coal (short ton)	6,287,000 Btu
Bituminous and lignite	21,514,000 Btu
Anthracite	22,435,000 Btu
Electricity (kilowatt-hour)	3,412 Btu
Natural gas liquids (barrel)	3,804,000 Btu

---

[*]These values were reported by the Energy Information Administration, U.S. Department of Energy, *Monthly Energy Review*, November 1988.

Appendix D

Principal Abbreviations and Acronyms

A

AFB	atmospheric fluidized bed
AGA	American Gas Association
AIChE	American Institute of Chemical Engineers
API	American Petroleum Institute
ASA	Aktiebolaget Svensk Alunskifferutveckilug (Sweden)
ASEAN	Association of Southeast Asian Nations
ASME	American Society of Mechanical Engineers
atm	atmosphere

B

bar	unit of pressure, 0.9868 atm or 10^5 Pascal
bbl, BBL	barrel
bcf, BCF	billion cubic feet
Bcm, BCM	billion cubic meter
BP	British Petroleum (Ltd)
Btu, BTU	British thermal unit

C

C	centigrade, Celsius
CFC	chlorofluorocarbon
Cif	cost, insurance and freight
CNG	compressed natural gas
COP	coefficient of performance
CRSTIG	chemically recuperated STIG

E

EEC	European Economic Community
EJ	exajoules
EPA	Environmental Protection Agency
EPRI	Electric Power Research Institute

F

FAO	Food and Agricultural Organization of the United Nations
FBC	fluid bed combustion
FERC	Federal Energy Regulatory Commission
FGD	fluid gas desulfurization
fob	free on board

G

GRI	Gas Research Institute

H

HRSG	heat recovery steam generator
HTS	high temperature shift reaction

I

ICI	Imperial Chemical Industries
IEA	International Energy Agency
IGCC	Integrated gasification combined cycle
IGU	International Gas Union
IRR	internal rate of return
ISTIG	intercooled steam-injected gas turbine

J

JANZ	Japan, Australia, New Zealand

K

kW	kilowatt
kWh	kilowatt-hour

L

LFG	landfill gas
LIMB	limestone injection multi-stage burner
LNG	liquified natural gas
LPG	liquified petroleum gas
LTS	low temperature shift reaction

M

Mcf, MCF	million cubic feet
MEA	monoethanolamine solution
MM	million
MM Btu	million Btu
MMT	million ton
MSW	municipal solid waste
MtBe	methyl t-butyl ether
MW	megawatt

N

N	nitrogen equivalent in nitrogenous materials
NGL	natural gas liquid
NGV	natural gas volatiles
NSPS	New Source Performance Standards

O

O&M	operation and maintenance
OECD	Organization for Economic Cooperation and Development
OTA	Office of Technology Assessment

P

pcd	pounds per capita per day (waste)
PFB	pressurized fluid bed
PM	particulate matter
PURPA	Public Utility Regulatory Policies Act (US)
PV	photovoltaic

Q

Q	10^{18} Btu; 10^{21} Joule
q	1 quad; 10^{15} Btu

R

RDF	refuse-derived fuel

S

SCR	selective catalytic reduction
SNG	substitute natural gas; snythetic natural gas
STAG	steam turbine and gas turbine combined cycle
STIG	steam-injected gas turbine

T

Tcf	trillion cubic feet
TFI	The Fertilizer Institute
tpd	tons per day (metric)
tpy	tons per year (metric)
TSP	total solids of particulates

U

UNIDO	United Nations Industrial Development Organization
USGS	United States Geological Survey

W

W	watt
WEC	World Energy Conference

ABOUT THE CONTRIBUTORS

Nelson Hay is Chief Economist and Director for Policy Analysis at the American Gas Association in Arlington, Va.

Afsaneh Mashayekhi is Chief of the Gas Unit with the Policy, Planning and Research Department of the World Bank in Washington, D.C.

Dr. Sergio Trindade is Assistant Secretary General at the United Nations and Director of the United Nations Center for Science and Technology in Developing Countries in New York.

Dr. Robert Williams and Dr. Eric Larson are with the School of Engineering/Applied Science, Center for Energy and Environmental Studies at Princeton University in Princeton, N.J.

William Sheldrick is the former Chairman of the FAO-UNIDO-World Bank roundtable on fertilizers and is currently a consultant based in Dorset, England.

Walter Vergara is a Chemical Engineer with the Asia Technical Department of the World Bank in Washington, D.C.

Robert Saunders and Rene Moreno, Jr. are Division Chief and Senior Economist, respectively, with the Industry and Energy Department at the World Bank in Washington, D.C.

Carl Hall is Deputy Assistant Director for Engineering at the National Science Foundation in Washington, D.C.

The Editors wish to thank Jennifer L. Nazak for her assistance with the layout of the book and the production of its camera-ready version.

Index

315

317

318